Lectures on Classical Differential Geometry

—— SECOND EDITION ——

Dirk J. Struik

MASSACHUSETTS INSTITUTE OF TECHNOLOGY

DOVER PUBLICATIONS, INC.
New York

Copyright © 1950, 1961 by Dirk J. Struik.

Published in Canada by General Publishing Company,
Ltd., 30 Lesmill Road, Don Mills, Toronto, Ontario.
Published in the United Kingdom by Constable and Com-
pany, Ltd.

This Dover edition, first published in 1988, is an un-
abridged and unaltered republication of the second edition
(1961) of the work first published in 1950 by the Addison-
Wesley Publishing Company, Inc., Reading, Massachusetts.

Manufactured in the United States of America
Dover Publications, Inc., 31 East 2nd Street, Mineola, N.Y.
11501

Library of Congress Cataloging-in-Publication Data

Struik, Dirk Jan, 1894–
    Lectures on classical differential geometry / Dirk J.
Struik. — 2nd ed.
        p.   cm.
    Reprint. Originally published: Reading, Mass. : Addi-
son-Wesley Pub. Co., 1961.
    Bibliography: p.
    Includes index.
    ISBN 0-486-65609-8
    1. Geometry, Differential.   I. Title.
QA641.S72   1988
516.3′602—dc19   87-34903
                                                      CIP

CONTENTS

PREFACE

This book has developed from a one-term course in differential geometry given for juniors, seniors, and graduate students at the Massachusetts Institute of Technology. It presents the fundamental conceptions of the theory of curves and surfaces and applies them to a number of examples. Some care is given to historical, biographical, and bibliographical material, not only to keep alive the memory of the men to whom we owe the main structure of our present elementary differential geometry, but also to allow the student to go back to the sources, which still contain many precious ideas for further thought.

No author on this subject is without primary obligation to the two standard treatises of Darboux and Bianchi, who, around the turn of the century, collected the result of more than a century of research, themselves adding greatly to it. Other fundamental works constantly consulted by the author are Eisenhart's *Treatise*, Scheffers' *Anwendung*, and Blaschke's *Vorlesungen*. Years of teaching from Graustein's *Differential Geometry* have also left their imprint on the presentation of the material.

The notation used is the Gibbs form of vector analysis, which after years of competition with other notations seems to have won the day, not only in the country of its inception, but also in many other parts of the world. Those unfamiliar with this notation may be aided by some explanatory remarks introduced in the text. This notation is amply sufficient for those more elementary aspects of differential geometry which form the subject of this introductory course; those who prefer to study our subject with tensor methods and thus to prepare themselves for more advanced research will find all they need in the books which Eisenhart and Hlavaty have devoted to this aspect of the theory. Some problems in the present book may serve as a preparation for this task.

Considerable attention has been paid to the illustrations, which may be helpful in stimulating the student's visual understanding of geometry. In the selection of his illustrations the author has occasionally taken his inspiration from some particularly striking pictures which have appeared in other books, or from mathematical models in the M.I.T. collection.* The

* In particular, the following figures have been wholly or in part suggested by other authors: Fig. 2-33 by Eisenhart, *Introduction;* Figs. 2-32, 2-33, 3-3, 5-14 by Scheffers, *Anwendung;* Figs. 2-22, 2-23, 5-7 by Hilbert-Cohn Vossen; and Fig. 5-9 by Adams (loc. cit., Section 5-3). The models, from L. Brill, Darmstadt, were constructed between 1877 and 1890 at the Universities of Munich (under A. Brill) and Göttingen (under H. A. Schwarz).

Art Department of Addison-Wesley Publishing Company is responsible for the excellent graphical interpretation and technique.

The problems in the text have been selected in such a way that most of them are simple enough for class use, at the same time often conveying an interesting geometrical fact. Some problems have been added at the end which are not all elementary, but reference to the literature may here be helpful to students ambitious enough to try those problems.

The author has to thank the publishers and their adviser, Dr. Eric Reissner, for their encouragement in writing this book and Mrs. Violet Haas for critical help. He owes much to the constructive criticism of his class in M 442 during the fall-winter term of 1949–50, which is responsible for many an improvement in text and in problems. He also acknowledges with appreciation discussions with Professor Philip Franklin, and the help of Mr. F. J. Navarro.

DIRK J. STRUIK

PREFACE TO THE SECOND EDITION

In this second edition some corrections have been made and an appendix has been added with a sketch of the application of Cartan's method of Pfaffians to curve and surface theory. This sketch is based on a paper presented to the sixth congress of the Mexican Mathematical Society, held at Merida in September, 1960.

A Spanish translation by L. Bravo Gala (Aguilar, Madrid, 1955) is only one of the many tokens that the book has been well received. The author likes to use this opportunity to express his appreciation and to thank those who orally or in writing have suggested improvements.

D.J.S.

BIBLIOGRAPHY

ENGLISH

EISENHART, L. P., *A treatise on the differential geometry of curves and surfaces.* Ginn & Co., Boston, etc., 1909, xi + 476 pp. (quoted as *Differential Geometry*).

EISENHART, L. P., *An introduction to differential geometry with use of the tensor calculus.* Princeton University Press, Princeton, 1940, 304 pp.

FORSYTH, A. R., *Lectures on the differential geometry of curves and surfaces.* University Press, Cambridge, 1912, 34 + 525 pp.

GRAUSTEIN, W. C., *Differential geometry.* Macmillan, New York, 1935, xi + 230 pp.

KREYSZIG, E., *Differential geometry.* University of Toronto Press, Toronto, 1959, xiv + 352 pp. Author's translation from the German: *Differentialgeometrie.* Akad. Verlagsges, Leipzig, 1957, xi + 421 pp.

LANE, E. P., *Metric differential geometry of curves and surfaces.* University of Chicago Press, Chicago, 1940, 216 pp.

POGORELOV, A. V., *Differential geometry.* Noordhoff, Groningen, 1959, ix + 171 pp. Translated from the Russian by L. F. BORON.

WEATHERBURN, C. E., *Differential geometry of three dimensions.* University Press, Cambridge, I, 1927, xii + 268 pp.; II, 1930, xii + 239 pp.

WILLMORE, T., *An introduction to differential geometry.* Clarendon Press, Oxford, 1959, 317 pp.

FRENCH

DARBOUX, G., *Leçons sur la théorie générale des surfaces.* 4 vols., Gauthier-Villars, Paris (2d ed., 1914), I, 1887, 513 pp.; II, 1889, 522 pp.; III, 1894, 512 pp.; IV, 1896, 548 pp. (quoted as *Leçons*).

FAVARD, J., *Cours de géométrie différentielle locale.* Gauthier-Villars, Paris, 1957, viii + 553 pp.

JULIA, G., *Elements de géométrie infinitésimale.* Gauthier-Villars, Paris, 2d ed., 1936, vii + 262 pp.

RAFFY, L., *Leçons sur les applications géométriques de l'analyse.* Gauthier-Villars, Paris, 1897, vi + 251 pp.

GERMAN

BLASCHKE, W., *Vorlesungen über Differentialgeometrie* I, 3d ed., Springer, Berlin, 1930, x + 311 pp. (quoted as *Differentialgeometrie*).

BLASCHKE, W., *Einführung in die Differentialgeometrie.* Springer-Verlag, 1950, vii + 146 pp.

HAACK, W., *Differential-geometrie*. Wolfenbüttler Verlagsanstalt. Wolfen-büttel-Hannover, 2 vols., 1948, I, 136 pp.; II, 131 pp.

HLAVATY, V., *Differentialgeometrie der Kurven und Flächen und Tensorrechnung*. Übersetzung von M. Pinl. Noordhoff, Groningen, 1939, xi + 569 pp.

KOMMERELL, V. UND K., *Allgemeine Theorie der Raumkurven und Flächen*. De Gruyter & Co., Berlin-Leipzig, 3ᵉ Aufl., 1921, I, viii + 184 pp.; II, 196 pp.

SCHEFFERS, G., *Anwendung der Differential- und Integralrechnung auf Geometrie*. De Gruyter, Berlin-Leipzig, 3ᵉ Aufl. I, 1923, x + 482 pp.; II, 1922, xi + 582 pp. (quoted as *Anwendung*).

STRUBECKER, K., *Differentialgeometrie*. Sammlung Göschen, De Gruyter, Berlin, I, 1955, 150 pp.; II, 1958, 193 pp.; III, 1959, 254 pp.

ITALIAN

BIANCHI, L., *Lezioni di geometria differenziale*. Spoerri, Pisa, 1894, viii + 541 pp.; 3ᵃ ed., I, 1922, iv + 806 pp.; II, 1923, 833 pp. German translation by M. LUKAT; *Vorlesungen über Differentialgeometrie*. Teubner, Leipzig, 2ᵉ Aufl., 1910, xviii + 721 pp. (Lezioni I quoted as *Lezioni*.)

RUSSIAN

FINIKOV, S. P., *Kurs differencial'noĭ geometrii*. Moscow, 1952, 343 pp.

RAŠEVSKIĬ, P. K., *Kurs differencial'noĭ geometrii*. Moscow, 4ᵗʰ ed., 1956, 420 pp.

VYGODSKII, M. YA., *Differential'naya geometriya*. Moscow-Leningrad, 1949, 511 pp.

There are also chapters on differential geometry in most textbooks of advanced calculus, such as:

GOURSAT, E., *Cours d'analyse mathématique*. Gauthier-Villars, Paris, I, 5ᵉ ed., 1943, 674 pp. English translation by E. R. HEDRICK, *Course in mathematical analysis*. Ginn & Co., Boston, I, 1904, viii + 548 pp.

The visual aspect of curve and surface theory is stressed in Chapter IV of D. HILBERT & S. COHN-VOSSEN, *Anschauliche Geometrie*. Springer, Berlin, 1932, viii + 310 pp.

The best collection of bibliographical notes and references in: *Encyklopädie der Mathematischen Wissenschaften*. Teubner, Leipzig, Band III, 3 Teil (1902–'27), 606 pp., article by H. V. MANGOLDT, R. V. LILIENTHAL, G. SCHEFFERS, A. VOSS, H. LIEBMANN, E. SALKOWSKI.

Also:

PASCAL, E., *Repertorium der Höheren Mathematik*. 2ᵉ Aufl. Zweiter Band, Teubner, Leipzig. Berlin, I (1910), II (1922), article by H. LIEBMANN and E. SALKOWSKI.

The history of differential geometry can be studied in COOLIDGE, J. L., *A history of geometrical methods*. Clarendon Press, Oxford, xviii + 451 pp., especially pp. 318–387.

Lectures on
Classical Differential
Geometry

CHAPTER 1

CURVES

1-1 Analytic representation. We can think of curves in space as paths of a point in motion. The rectangular coordinates (x, y, z) of the point can then be expressed as functions of a parameter u inside a certain closed interval:

$$x = x(u), \quad y = y(u), \quad z = z(u); \qquad u_1 \leqslant u \leqslant u_2. \tag{1-1a}$$

It is often convenient to think of u as the time, but this is not necessary, since we can pass from one parameter to another by a substitution $u = f(v)$ without changing the curve itself. We select the coordinate axes in such a way that the sense $OX \to OY \to OZ$ is that of a right-handed screw. We also denote (x, y, z) by (x_1, x_2, x_3), or for short, $x_i, i = 1, 2, 3$. The equation of the curve then takes the form

$$x_i = x_i(u); \qquad u_1 \leqslant u \leqslant u_2. \tag{1-1b}$$

We use the notation $P(x_i)$ to indicate a point with coordinates x_i.

EXAMPLES. (1) *Straight line.* A straight line in space can be given by the equation

$$x_i = a_i + ub_i, \tag{1-2}$$

where a_i, b_i are constants and at least one of the $b_i \neq 0$.

This equation represents a line passing through the point (a_i) with its direction cosines proportional to b_i. Eq. (1-2) can also be written:

$$\frac{x_1 - a_1}{b_1} = \frac{x_2 - a_2}{b_2} = \frac{x_3 - a_3}{b_3}.$$

(2) *Circle.* The circle is a plane curve. Its plane can be taken as $z = 0$ and its equation can then be written in the form:

$$x = a \cos u, \quad y = a \sin u, \quad z = 0; \\ 0 \leqslant u < 2\pi. \tag{1-3}$$

Here a is the radius, $u = \angle POX$ (Fig. 1-1).

(3) *Circular helix.* The equation is

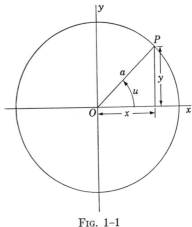

Fig. 1-1

1

$$x = a \cos u, \quad y = a \sin u, \quad z = bu; \qquad a, b \text{ constants.} \qquad (1\text{--}4)$$

This curve lies on the cylinder $x^2 + y^2 = a^2$ and winds around it in such a way that when u increases by 2π the x and y return to their original value, while z increases by $2\pi b$, the *pitch* of the helix (French: *pas;* German: *Ganghöhe*). When b is positive the helix is *right-handed* (Fig. 1–2a); when b is negative it is *left-handed* (Fig. 1–2b). This sense of the helix is independent of the choice of coordinates or parameters; it is an *intrinsic* property of the helix. A left-handed helix can never be superimposed on a right-handed one, as everyone knows who has handled screws or ropes.

The functions $x_i(u)$ are not all constants. If two of them are constants Eqs. (1–1) represent a straight line parallel to a coordinate axis. We also suppose that in the given interval of u the functions $x_i(u)$ are single-valued and continuous, with a sufficient number of continuous derivatives (first derivatives in all cases, seldom more than three). It is sufficient for this purpose to postulate that there exists at a point P of the curve, where

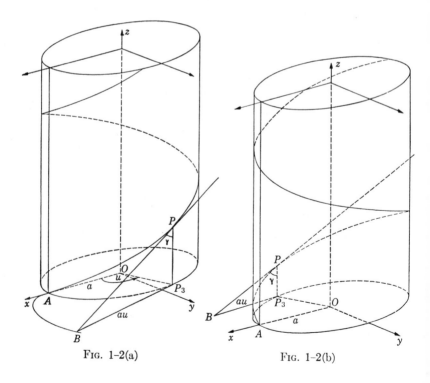

Fig. 1–2(a)   Fig. 1–2(b)

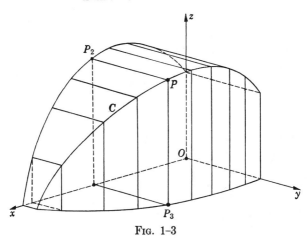

Fig. 1-3

$u = u_0$, a set of finite derivatives $x_i^{(n+1)}(u_0)$, n sufficiently large. Then we can express $x_i(u_0 + h)$ as follows in a Taylor development:

$$x_i(u_0 + h) = x_i(u)$$
$$= x_i(u_0) + \frac{h}{1}\,\dot{x}_i(u_0) + \frac{h^2}{2!}\,\ddot{x}_i(u_0) + \cdots + \frac{h^n}{n!}\,x_i^{(n)}(u_0) + o(h^n), \quad (1-5)$$

where the $\dot{x}_i$, $\ddot{x}_i$, ... $x_i^{(n)}$ represent derivatives with respect to u and $o(h^n)$ is a term such that

$$\lim_{h \to 0} \frac{o(h^n)}{h^n} = 0.^*$$

This is always satisfied, for all values of n, $n > 0$, when the $x_i(u)$ are complex functions of a complex u and the first derivatives $\dot{x}_i$ exist. The functions $x_i(u)$ are then *analytic*. *However, we usually consider the x_i as real functions of a real variable u.* The curve (1-1) with the conditions (1-5) is better called an *arc of curve*, but we shall continue to use the term *curve* as long as no ambiguity occurs. Points where all $\dot{x}_i$ vanish are called *singular*, otherwise *regular* with respect to u. When speaking of points, we mean regular points. When we replace the parameter u by another parameter,

$$u = f(u_1), \qquad (1-6)$$

---

* See e.g. P. Franklin, *A treatise on advanced calculus*, John Wiley and Sons, New York, 1940, p. 127; Ch. J. de la Vallée Poussin, *Cours d'analyse infinitésimale*, Dover Publications, New York, 1946, Tome I, 8th ed., p. 80.

we postulate that $f(u_1)$ be differentiable; when $du/du_1 \neq 0$ regular points remain regular.

Curves can also be defined in ways different from (1–1). We can use the equations

$$F_1(x, y, z) = 0, \qquad F_2(x, y, z) = 0, \tag{1–7}$$

or

$$y = f_1(x), \qquad z = f_2(x), \tag{1–8}$$

to define a curve. The type (1–8) can be considered as a special form of (1–1), x being taken as parameter. We obtain it from (1–1) by eliminating u from y and x, and also from z and x. This is always possible when $dx/du \neq 0$, so that u can be expressed in x. Type (1–8) expresses the curve C as the intersection of two projecting cylinders (Fig. 1–3). As to Eqs. (1–7), they define two implicit functions $y(x)$ and $z(x)$ when the functional determinant

$$\binom{F_1 F_2}{y \; z} \equiv \frac{\partial F_1}{\partial z} \frac{\partial F_2}{\partial y} - \frac{\partial F_2}{\partial z} \frac{\partial F_1}{\partial y} \neq 0.^*$$

This brings us to (1–8) and thus to (1–1).

The representations (1–7) and (1–8) define the space curve as the intersection of two surfaces. But such an intersection may split into several curves. If, for instance, F_1 and F_2 represent two cones with a common generating line, Eqs. (1–7) define this line together with the remaining intersection. And if we eliminate u from the equations

$$x = u, \qquad y = u^2, \qquad z = u^3, \quad (1–9)$$

which represent a space curve C of the third degree (a *cubic parabola*), we obtain the equations

$$y = x^2, \qquad xz - y^2 = 0,$$

which represent the intersection of a cylinder (Fig. 1–4) and a cone. This intersection contains not only the curve (1–9), but also the Z-axis.

The complete intersection of an algebraic surface of degree m and an algebraic surface of degree n is a space curve of degree mn, which may

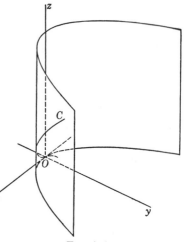

Fig. 1–4

---

* Franklin, loc. cit., pp. 340–341.

split into several curves, the degree of which adds up to mn. We say that a surface is of degree k when it is intersected by a line in k points or, what is the same, by a plane in a curve of degree k. A space curve is of degree l if a plane intersects it in l points. The points of intersection may be real, imaginary, coincident, or at infinity. In the case (1–9) we substitute x, y, z into the equation of a plane $ax + by + cz + d = 0$, and obtain a cubic equation for u, which has three roots, indicating three points of intersection.

This explains why we often prefer to give a curve by equations of the form (1–1). Moreover, this presentation allows a ready application of the ideas of vector analysis.

For this purpose, let $\mathbf{e}_1$, $\mathbf{e}_2$, $\mathbf{e}_3$, or for short $\mathbf{e}_i$ $(i = 1, 2, 3)$ be unit vectors in the direction of the positive X, Y, and Z-axes. Then we can give a curve C by expressing the radius vector $\overrightarrow{OP} = \mathbf{x}$ of a generic point P as a function of u (Fig. 1–5) in the following way:

$$\mathbf{x} = x_1\mathbf{e}_1 + x_2\mathbf{e}_2 + x_3\mathbf{e}_3, \quad (1\text{–}10)$$

where the x_i are given by Eq. (1–1). We indicate P not only by $P(x_i)$, but also by $P(\mathbf{x})$ or $P(u)$; we shall also speak of "the curve $\mathbf{x}(u)$."

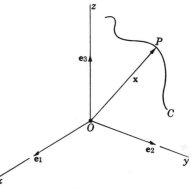

Fig. 1–5

The x_i are the *coordinates* of $\mathbf{x}$, the vectors $x_1\mathbf{e}_1, x_2\mathbf{e}_2, x_3\mathbf{e}_3$ are the *components* of $\mathbf{x}$ along the coordinate axes. We shall often indicate a vector by its coordinates, as $\mathbf{x}(x, y, z)$, or as $\mathbf{x}(x_i)$.

The length of the (real) vector $\mathbf{x}$ is indicated by

$$|\mathbf{x}| = \sqrt{x_1^2 + x_2^2 + x_3^2}. \quad (1\text{–}11)$$

1–2 Arc length, tangent. We suppose in this and in the next sections (until Sec. 1–12) that the curve C is real, with real u. Then, as shown in the texts on calculus,[*] we can express the arc length of a segment of the curve between points $A(u_0)$ and $P(u)$ by means of the integral

$$s(u) = \int_{u_0}^{u} \sqrt{\dot{x}^2 + \dot{y}^2 + \dot{z}^2}\, du = \int_{u_0}^{u} \sqrt{\dot{\mathbf{x}} \cdot \dot{\mathbf{x}}}\, du. \quad (2\text{–}1)$$

---

[*] See e.g. P. Franklin, *A treatise on advanced calculus*, John Wiley and Sons, New York, 1940, pp. 284, 294; Ch. J. de la Vallée Poussin, *Cours d'analyse infinitésimale*, Dover Publications, New York, 1946, Tome I, 8th ed., p. 272.

The dot will always indicate differentiation with respect to u:

$$\dot{\mathbf{x}} = d\mathbf{x}/du, \qquad \dot{x}_i = dx_i/du. \tag{2-2}$$

The square root is positive. The expression $\dot{\mathbf{x}} \cdot \dot{\mathbf{x}}$ is the scalar product of $\dot{\mathbf{x}}$ with itself; it is always > 0 for real curves. $\dot{\mathbf{x}}$ is assumed to be nowhere zero in the given interval (no singular points, see Sec. 1–1).

We define the scalar product of two vectors $\mathbf{v}(v_i)$ and $\mathbf{w}(w_i)$ by the formula:

$$\mathbf{v} \cdot \mathbf{w} = \mathbf{w} \cdot \mathbf{v} = v_1 w_1 + v_2 w_2 + v_3 w_3. \tag{2-3}$$

It can be shown that, φ being the angle between $\mathbf{v}$ and $\mathbf{w}$ (Fig. 1–6),

$$\mathbf{v} \cdot \mathbf{w} = |\mathbf{v}||\mathbf{w}| \cos \varphi. \tag{2-4}$$

This shows that $\mathbf{v} \cdot \mathbf{w} = 0$ means that $\mathbf{v}$ and $\mathbf{w}$ are perpendicular. A unit vector $\mathbf{u}$ satisfies the equations

$$|\mathbf{u}| = 1, \qquad \mathbf{u} \cdot \mathbf{u} = 1.$$

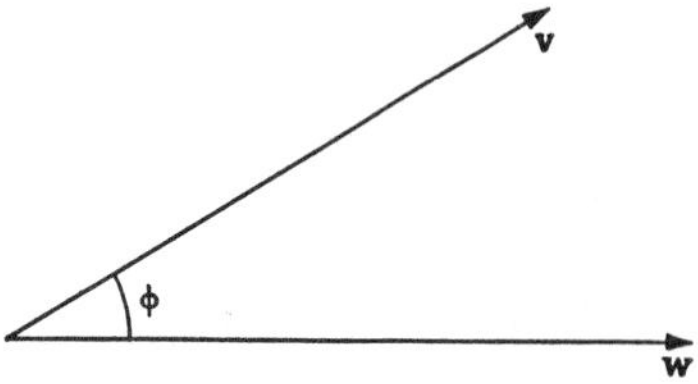

Fig. 1–6

The arc length s increases with increasing u. The sense of increasing arc length is called the *positive* sense on the curve; a curve with a sense on it is called an *oriented* curve. Most of our reasoning in differential geometry is with oriented curves; our space has also been oriented by the introduction of a right-handed coordinate system. However, our results are often independent of the orientation.

When we change the parameter on the curve from u to u_1 the arc length retains its form, with u_1 instead of u. We can express this *invariance* under parameter transformations by replacing Eq. (2–1) with the equation

$$ds^2 = dx^2 + dy^2 + dz^2 = d\mathbf{x} \cdot d\mathbf{x}, \tag{2-5}$$

which is independent of u.

When we now introduce s as parameter instead of u — which is always legitimate, since $ds/du \neq 0$ — then Eq. (2–5) shows us that

$$\frac{d\mathbf{x}}{ds} \cdot \frac{d\mathbf{x}}{ds} = 1. \tag{2-6}$$

The vector $d\mathbf{x}/ds$ is therefore a unit vector. It has a simple geometrical interpretation. The vector $\Delta\mathbf{x}$ joins two points $P(\mathbf{x})$ and $Q(\mathbf{x} + \Delta\mathbf{x})$ on the curve. The vector $\Delta\mathbf{x}/\Delta s$ has the same direction as $\Delta\mathbf{x}$ and for $\Delta s \to 0$

passes into a tangent vector at P (Fig. 1–7). Since its length is 1 we call the vector

$$\mathbf{t} = d\mathbf{x}/ds \qquad (2\text{–}7)$$

the *unit tangent vector* to the curve at P. Its sense is that of increasing s. Since

$$\frac{d\mathbf{x}}{du} = \frac{d\mathbf{x}}{ds}\frac{ds}{du}, \qquad (2\text{–}8)$$

we see that $\dot{\mathbf{x}} = d\mathbf{x}/du$ is also a tangent vector, though not necessarily a unit vector.

We often express the fact that the tangent is the limiting position of a line through P and a point Q in the given interval of u, when $Q \to P$, by saying that *the tangent passes through two consecutive points on the curve.* This mode of expression seems unsatisfactory, but it has considerable heuristic value and can still be made quite rigorous.

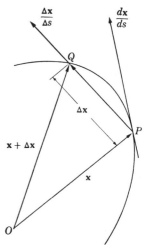

Fig. 1–7

THEOREM. *The ratio of the arc and the chord connecting two points P and Q on a curve approaches unity when Q approaches P.*

Indeed, when Δs is the arc PQ and c is the chord PQ, then for $Q \to P$ (Fig. 1–8):

$$\lim \frac{\Delta s}{c} = \lim \frac{\Delta s}{\sqrt{(\Delta x)^2 + (\Delta y)^2 + (\Delta z)^2}}$$

$$= \lim \frac{\Delta s/\Delta u}{\sqrt{\left(\dfrac{\Delta x}{\Delta u}\right)^2 + \left(\dfrac{\Delta y}{\Delta u}\right)^2 + \left(\dfrac{\Delta z}{\Delta u}\right)^2}}$$

$$= \frac{\dot{s}}{\sqrt{\dot{x}^2 + \dot{y}^2 + \dot{z}^2}} = 1, \qquad (2\text{–}9)$$

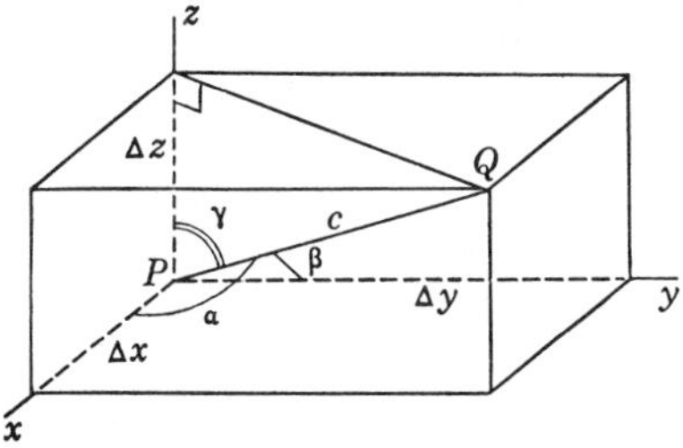

Fig. 1–8

which proves the theorem. It also proves that the ratio of Δu and c is finite.

The ratios $\Delta x/\Delta s$, $\Delta y/\Delta s$, $\Delta z/\Delta s$ (Fig. 1–8) therefore approach, for $\Delta s \to 0$, the cosines of the angles which the oriented tangent at P makes with the positive X-, Y-, and Z-axes. This means that $\mathbf{t}$ can be expressed in the form:

$$\mathbf{t} = \mathbf{e}_1 \cos \alpha_1 + \mathbf{e}_2 \cos \alpha_2 + \mathbf{e}_3 \cos \alpha_3, \qquad (2\text{–}10)$$

which is in accordance with the identity

$$\cos^2 \alpha_1 + \cos^2 \alpha_2 + \cos^2 \alpha_3 = 1. \qquad (2\text{--}11)$$

A generic point $A(\mathbf{X})$ on the tangent line at P is determined by the equation (Fig. 1–9):

$$\mathbf{X} = \mathbf{x} + v\mathbf{t}, \qquad v = PA, \qquad (2\text{--}12)$$

in coordinates (supposing all $dx_i \neq 0$)

$$\frac{X_1 - x_1}{\cos \alpha_1} = \frac{X_2 - x_2}{\cos \alpha_2} = \frac{X_3 - x_3}{\cos \alpha_3} \qquad (2\text{--}13)$$

or

$$\frac{X_1 - x_1}{dx_1} = \frac{X_2 - x_2}{dx_2} = \frac{X_3 - x_3}{dx_3}. \qquad (2\text{--}14)$$

EXAMPLES. (1) *Circle* (Fig. 1–10).

$$x = a \cos u, \qquad y = a \sin u, \qquad z = 0; \qquad (2\text{--}15)$$
$$\dot{x} = -a \sin u, \qquad \dot{y} = a \cos u;$$
$$s = au + \text{const}, \qquad \text{take } s = au;$$
$$x = a \cos (s/a), \qquad y = a \sin (s/a).$$

Unit tangent vector: $\mathbf{t}(-\sin u, \cos u)$.

Equation of tangent line:

$$\frac{X_1 - a \cos u}{-\sin u} = \frac{X_2 - a \sin u}{\cos u},$$

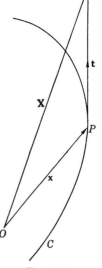

FIG. 1–9

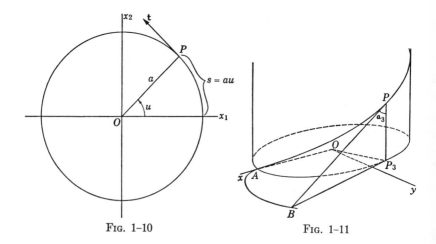

FIG. 1–10

FIG. 1–11

or, writing for (X_1X_2) again (xy):

$$x \cos u + y \sin u = a.$$

(2) *Circular helix* (Fig. 1–11).

$$x = a \cos u, \qquad y = a \sin u, \qquad z = bu; \qquad (2\text{–}16)$$
$$\dot{x} = -a \sin u, \qquad \dot{y} = a \cos u, \qquad \dot{z} = b, \qquad s = u\sqrt{a^2 + b^2} = cu$$

$$\mathbf{t}\left(-\frac{a}{c} \sin u, \frac{a}{c} \cos u, \frac{b}{c}\right).$$

The tangent vector makes a constant angle α_3 with the Z-axis:

$$\cos \alpha_3 = b/c, \qquad \text{hence } \tan \alpha_3 = a/b.$$

If B is the intersection of the tangent at P with the XOY-plane, and P_3 the projection of P on this plane, then

$$P_3B = PP_3 \tan \alpha_3 = bu\frac{a}{b} = au = \text{arc } AP_3.$$

The locus of B is therefore the involute of the basic circle of the cylinder (see Sec. 1–11).

(3) *A space curve of degree four* (Fig. 1–12).

$$x = a(1 + \cos u), \qquad y = a \sin u,$$
$$z = 2a \sin u/2.$$

For $0 < u < 2\pi$ we obtain the points above the XOY-plane; for $-2\pi < u < 0$ those below this plane. The whole curve is described for $-2\pi < u < 2\pi$. Elimination of u gives

$$x^2 + y^2 + z^2 = 4a^2,$$
$$(x - a)^2 + y^2 = a^2.$$

The curve is the full intersection of the sphere with center at O and radius $2a$, and the circular cylinder with radius a and axis in the XOZ-plane

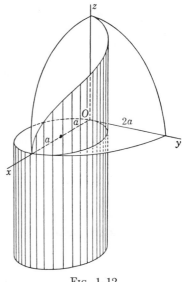

Fig. 1–12

at distance a from OZ. The substitution $u = \tan u/4$ allows us to write the coordinates of the curve by means of rational functions:

$$x = \frac{2a(1 - u^2)^2}{(1 + u^2)^2}, \qquad y = \frac{4au(1 - u^2)}{(1 + u^2)^2}, \qquad z = \frac{4au}{1 + u^2}.$$

Substitution of these expressions into the equation of a plane,

$$ax + by + cz + d = 0,$$

gives a biquadratic equation, showing that the degree of the curve is four indeed. The arc length

$$s = a \int_0^u \sqrt{1 + \cos^2 \frac{u}{2}} \, du$$

is an elliptic integral.

The reader will recognize in Fig. 1–12 the so-called temple of Viviani (1692), well known in the theory of multiple integration, remarkable because both the area and the volume of the hemisphere $x > 0$ outside the cylinder is rational in a.

1–3 Osculating plane. The tangent can be defined as the line passing through two consecutive points of the curve. We shall now try to find the plane through three consecutive points, which means the limiting position of a plane passing through three nearby points of the curve when two of these points approach the third. For this purpose let us consider a plane

$$\mathbf{X} \cdot \mathbf{a} = p; \quad \mathbf{X} \text{ generic point of the plane, } \mathbf{a} \perp \text{plane, } p \text{ a constant,} \quad (3\text{–}1)$$

passing through the points P, Q, R on the curve given by $\mathbf{X} = \mathbf{x}(u_0)$, $\mathbf{X} = \mathbf{x}(u_1)$, $\mathbf{X} = \mathbf{x}(u_2)$. Then the function

$$f(u) = \mathbf{x} \cdot \mathbf{a} - p, \qquad \mathbf{x} = \mathbf{x}(u) \qquad (3\text{–}2)$$

satisfies the conditions

$$f(u_0) = 0, \qquad f(u_1) = 0, \qquad f(u_2) = 0.$$

Hence, according to Rolle's theorem, there exist the relations

$$f'(v_1) = 0, \qquad f'(v_2) = 0, \qquad u_0 \leqslant v_1 \leqslant u_1, \qquad u_1 \leqslant v_2 \leqslant u_2,$$

and

$$f''(v_3) = 0, \qquad v_1 \leqslant v_3 \leqslant v_2.$$

When Q and R approach P, $u_1, u_2, v_1, v_2, v_3 \to u_0$, and writing u for u_0, we obtain, for the limiting values of $\mathbf{a}$ and p, the conditions:

$$\begin{aligned} f(u) &= \mathbf{x} \cdot \mathbf{a} - p = 0, \\ f'(u) &= \dot{\mathbf{x}} \cdot \mathbf{a} = 0, \qquad\qquad (\dot{\mathbf{x}}, \ddot{\mathbf{x}} \text{ at } P(u)) \quad (3\text{–}3) \\ f''(u) &= \ddot{\mathbf{x}} \cdot \mathbf{a} = 0. \end{aligned}$$

Eliminating $\mathbf{a}$ from Eqs. (3–3) and (3–1) we obtain a linear relation between $\mathbf{X} - \mathbf{x}, \dot{\mathbf{x}}, \ddot{\mathbf{x}}$ (all $\perp \mathbf{a}$):

$$\mathbf{X} = \mathbf{x} + \lambda\dot{\mathbf{x}} + \mu\ddot{\mathbf{x}}, \quad \lambda, \mu \text{ arbitrary constants}, \quad (3\text{–}4a)$$

in coordinates

$$\begin{vmatrix} X_1 - x_1 & X_2 - x_2 & X_3 - x_3 \\ \dot{x}_1 & \dot{x}_2 & \dot{x}_3 \\ \ddot{x}_1 & \ddot{x}_2 & \ddot{x}_3 \end{vmatrix} = 0 \qquad (3\text{–}4b)$$

This is the equation of the plane through three consecutive points of the curve, to which John Bernoulli has given the pleasant name of *osculating plane* (German: *Schmiegungsebene*). *It passes through* (at least) *three consecutive points of the curve.* It also passes through the tangent line (given by $\mu = 0$ in Eq. (3–4a)). The osculating plane is not determined when $\ddot{\mathbf{x}} = 0$ or when $\ddot{\mathbf{x}}$ is proportional to $\dot{\mathbf{x}}$.

We can express Eqs. (3–4) in another way by the introduction of the *vector product* of two vectors $\mathbf{v}$ and $\mathbf{w}$:

$$\mathbf{v} \times \mathbf{w} = \begin{vmatrix} \mathbf{e}_1 & \mathbf{e}_2 & \mathbf{e}_3 \\ v_1 & v_2 & v_3 \\ w_1 & w_2 & w_3 \end{vmatrix} = -\mathbf{w} \times \mathbf{v}. \qquad (3\text{–}5)$$

It can be shown that, φ being the angle of $\mathbf{v}$ and $\mathbf{w}$,

$$\mathbf{v} \times \mathbf{w} = |\mathbf{v}||\mathbf{w}| \sin \varphi \, \mathbf{u}, \qquad (3\text{–}6)$$

where $\mathbf{u}$ is a unit vector perpendicular to $\mathbf{v}$ and $\mathbf{w}$ in such a way that the sense $\mathbf{v} \to \mathbf{w} \to \mathbf{u}$ is the same as that of $OX \to OY \to OZ$, hence a right-handed one. We always have $\mathbf{v} \times \mathbf{v} = 0$. When $\mathbf{v} \times \mathbf{w} = 0$, then $\mathbf{w}$ has the direction of $\mathbf{v}$, or $\mathbf{w} = \lambda\mathbf{v}$; in this case we say that $\mathbf{v}$ and $\mathbf{w}$ are *collinear*.

The *triple scalar product* (or *parallelepiped product*) of three vectors $(\mathbf{v}, \mathbf{w}, \mathbf{u})$ is:

$$(\mathbf{v} \times \mathbf{w}) \cdot \mathbf{u} = \begin{vmatrix} u_1 & u_2 & u_3 \\ v_1 & v_2 & v_3 \\ w_1 & w_2 & w_3 \end{vmatrix} = \mathbf{u} \cdot (\mathbf{v} \times \mathbf{w}) = \mathbf{v} \cdot (\mathbf{w} \times \mathbf{u}) = \text{etc.}$$
$$= (\mathbf{vwu}) = (\mathbf{wuv}) = (\mathbf{uvw}) = \text{etc.} \qquad (3\text{–}7)$$

It is zero when the three vectors (each supposedly $\neq 0$) are *coplanar*, that is, can be moved into one plane.

The following formula is also useful:

$$(\mathbf{a} \times \mathbf{b}) \cdot (\mathbf{c} \times \mathbf{d}) = (\mathbf{a} \cdot \mathbf{c})(\mathbf{b} \cdot \mathbf{d}) - (\mathbf{a} \cdot \mathbf{d})(\mathbf{b} \cdot \mathbf{c}), \qquad (3\text{–}8)$$

with the special case

$$(\mathbf{a} \times \mathbf{b}) \cdot (\mathbf{a} \times \mathbf{b}) = (\mathbf{a} \cdot \mathbf{a})(\mathbf{b} \cdot \mathbf{b}) - (\mathbf{a} \cdot \mathbf{b})^2. \qquad (3\text{–}9)$$

The equation of the osculating plane can also be written (comp. Eq. (3–7) with Eq. (3–4b)):

$$(\mathbf{X} - \mathbf{x}, \dot{\mathbf{x}}, \ddot{\mathbf{x}}) = 0. \tag{3–10}$$

As to the two exceptional cases, if they are valid at all points of the curve then they both are satisfied for straight lines and only for those:

(a) $\ddot{\mathbf{x}} = 0$, $\qquad \dot{\mathbf{x}} = \mathbf{a}$, $\qquad\qquad \mathbf{x} = u\mathbf{a} + \mathbf{b}$

(b) $\ddot{\mathbf{x}} = \lambda\dot{\mathbf{x}}$, $\qquad \dot{\mathbf{x}} = \mathbf{c}e^{\lambda t} = \mathbf{c}f_1(t)$, $\qquad \mathbf{x} = \mathbf{c}f_2(t) + \mathbf{d}$,

where $\mathbf{a}$, $\mathbf{b}$, $\mathbf{c}$, $\mathbf{d}$ are fixed vectors.

If $\ddot{\mathbf{x}} = \lambda\dot{\mathbf{x}}$ ($\ddot{\mathbf{x}} \neq 0$) at one point of the curve, then we call this point a *point of inflection*. The tangent at such a point has three consecutive points in common with the curve (see Section 1–7).

Since PQ passes into the tangent at P, and QR into the tangent at Q, we say that *the osculating plane contains two consecutive tangent lines.* This indicates that we may facilitate our understanding of the nature of the osculating planes by taking (Fig. 1–13) the points $P_1, P_2, P_3, \ldots$ on the curve and considering the polygonal line $P_1P_2P_3 \ldots$ The sides $P_1P_2, P_2P_3, \ldots$ are all very short, and represent the tangent lines; the planes $P_1P_2P_3$, $P_2P_3P_4, \ldots$ represent the osculating planes. Two consecutive tangent lines P_1P_2, P_2P_3 lie in the osculating plane $P_1P_2P_3$; two consecutive osculating planes $P_1P_2P_3$, $P_2P_3P_4$ intersect in the tangent line P_2P_3, etc.

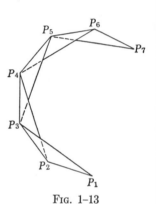

Fig. 1–13

EXAMPLES. (1) *Plane curve.* Since in this case the lines PQ and QR, considered in Eq. (3–1), lie in the plane of the curve, the osculating plane coincides with the plane of the curve. This is also clear from Fig. 1–13. When the curve is a straight line the osculating plane is indeterminate and may be any plane through the line.

(2) *Circular helix.* The equation of the osculating plane is

$$\begin{vmatrix} X_1 - a\cos u & X_2 - a\sin u & X_3 - bu \\ -a\sin u & a\cos u & b \\ -a\cos u & -a\sin u & 0 \end{vmatrix} = 0, \tag{3–10}$$

or, writing (x, y, z) for (X_1, X_2, X_3):

$$bx\sin u - by\cos u + az = abu.$$

This equation is satisfied by $x = \lambda \cos u$, $y = \lambda \sin u$, $z = bu$ for all values of λ, which (by fixed u) shows that the osculating plane at P contains the line PA parallel to the XOY-plane intersecting the Z-axis. The plane through PA and the tangent at P is the osculating plane at P (Fig. 1–14). The locus of the lines PA, indicated by $P_1A_1, P_2A_2, P_3A_3, \ldots$, is a surface called the *right helicoid* (see Sec. 2–2, 2–8).

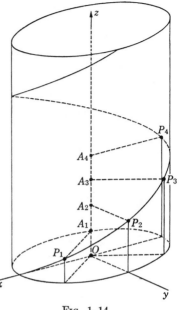

Fig. 1–14

1–4 Curvature. The line in the osculating plane at P perpendicular to the tangent line is called the *principal normal* (e.g., the lines AP in Fig. 1–14). In its direction we place a unit vector $\mathbf{n}$, the sense of which may be arbitrarily selected, provided it is continuous along the curve. If we now take the arc length as parameter:

$$\mathbf{x} = \mathbf{x}(s), \qquad \mathbf{t} = d\mathbf{x}/ds = \mathbf{x}', \qquad \mathbf{t} \cdot \mathbf{t} = 1, \qquad (4-1)$$

where the prime signifies differentiation with respect to s, then we obtain by differentiating $\mathbf{t} \cdot \mathbf{t} = 1$:

$$\mathbf{t} \cdot \mathbf{t}' = 0. \qquad (4-2)$$

This shows that the vector $\mathbf{t}' = d\mathbf{t}/ds$ is perpendicular to $\mathbf{t}$, and since

$$\mathbf{t} = \mathbf{x}' = \dot{\mathbf{x}}u', \qquad \mathbf{t}' = \ddot{\mathbf{x}}(u')^2 + \dot{\mathbf{x}}u'', \qquad (4-3)$$

we see that $\mathbf{t}'$ lies in the plane of $\dot{\mathbf{x}}$ and $\ddot{\mathbf{x}}$, and hence in the osculating plane. We can therefore introduce a proportionality factor κ such that

$$\mathbf{k} = d\mathbf{t}/ds = \kappa\mathbf{n}. \qquad (4-4)$$

The vector $\mathbf{k} = d\mathbf{t}/ds$, which expresses the rate of change of the tangent when we proceed along the curve, is called the *curvature vector*. The factor κ is called the *curvature*; $|\kappa|$ is the length of the curvature vector. Although the sense of $\mathbf{n}$ may be arbitrarily chosen, that of $d\mathbf{t}/ds$ is perfectly determined by the curve, independent of its orientation; when s changes sign, $\mathbf{t}$

also changes sign.   When **n** (as is often done) is taken in the sense of dt/ds, then κ is always positive, but we shall not adhere to this convention (see below, small type).

When we compare the tangent vectors $\mathbf{t}(u)$ at P and $\mathbf{t} + \Delta\mathbf{t}(u + h)$ at Q (Fig. 1–15) by moving $\mathbf{t}$ from P to Q, then $\mathbf{t}$, $\Delta\mathbf{t}$ and $\mathbf{t} + \Delta\mathbf{t}$ form an isosceles triangle with two sides equal to 1, enclosing the angle $\Delta\varphi$, the *angle of contingency*.   Since

$$|\Delta\mathbf{t}| = 2 \sin \Delta\varphi/2$$
$$= \Delta\varphi + \text{terms of higher order in } \Delta\varphi,$$

we find for $\Delta\varphi \to 0$:

$$|\kappa| = |dt/ds| = |\mathbf{k}| = |d\varphi/ds|, \qquad (4\text{–}5)$$

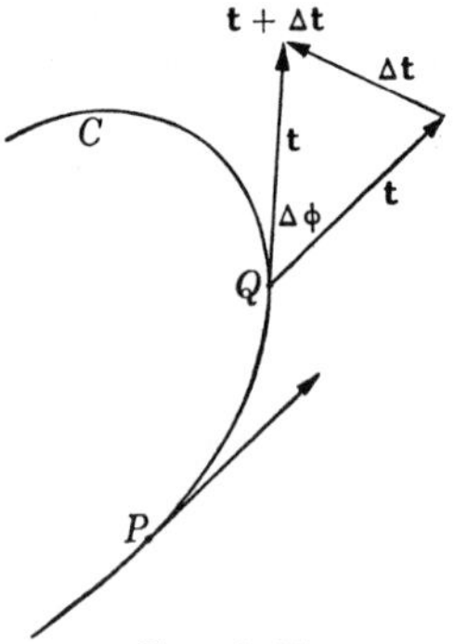

Fig. 1–15

which is the usual definition of the curvature in the case of a plane curve.

From Eq. (4–4) follows:

$$\kappa^2 = \mathbf{x}'' \cdot \mathbf{x}''. \qquad (4\text{–}6)$$

We define R as κ^{-1}.   *The absolute value of R is the radius of curvature, which is the radius of the circle passing through three consecutive points of the curve, the osculating circle.*   To prove it, we first observe that this circle lies in the osculating plane.   Let a circle be determined in this plane as intersection of the plane and the sphere given by

$$(\mathbf{X} - \mathbf{c}) \cdot (\mathbf{X} - \mathbf{c}) - r^2 = 0 \quad (\mathbf{X} \text{ generic point of the sphere, } \mathbf{c} \text{ center, } r \text{ radius}).$$

This sphere must pass through the points P, Q, R on the curve given by $\mathbf{X} = \mathbf{x}(s_0)$, $\mathbf{X} = \mathbf{x}(s_1)$, $\mathbf{X} = \mathbf{x}(s_2)$; the vector $\mathbf{c}$ points from O to a point in the osculating plane so that $\mathbf{x} - \mathbf{c}$ lies in the osculating plane.   Reasoning as in Sec. 1–3 on the function

$$f(s) = (\mathbf{x} - \mathbf{c}) \cdot (\mathbf{x} - \mathbf{c}) - r^2, \qquad \mathbf{x} = \mathbf{x}(s), \mathbf{c}, r \text{ constants},$$

we find, for the limiting values of $\mathbf{c}$ and r, the conditions

$$\begin{aligned}
f(s) &= 0, \\
f'(s) &= 0, \quad \text{or} \quad (\mathbf{x} - \mathbf{c}) \cdot \mathbf{x}' = 0, \\
f''(s) &= 0, \quad \text{or} \quad (\mathbf{x} - \mathbf{c}) \cdot \mathbf{x}'' + \mathbf{x}' \cdot \mathbf{x}' = (\mathbf{x} - \mathbf{c}) \cdot \mathbf{x}'' + 1 = 0.
\end{aligned} \qquad (4\text{–}7)$$

Since $\mathbf{x} - \mathbf{c}$ lies in the osculating plane, it is a linear combination of $\mathbf{x}'$ and $\mathbf{x}''$.   Hence

$$\mathbf{x} - \mathbf{c} = \lambda\mathbf{x}' + \mu\mathbf{x}'',$$

where λ and μ are determined by Eq. (4–7). We find $\lambda = 0, -1 = \mu \mathbf{x}'' \cdot \mathbf{x}''$, so that

$$\mathbf{c} = \mathbf{x} - \mu \mathbf{x}'',$$

or, in consequence of Eqs. (4–4) and (4–6):

$$\mathbf{c} = \mathbf{x} + \kappa \mathbf{n}/\kappa^2 = \mathbf{x} + R\mathbf{n}. \tag{4–8}$$

This shows that *the center of the osculating circle lies on the principal normal* at distance $|R|$ from P. Though $R = \kappa^{-1}$ may be positive or negative, the vector $R\mathbf{n}$ is independent of the sense of $\mathbf{n}$, having the sense of the curvature vector. Its end point is also called the center of curvature.

Eq. (4–6) shows algebraically that the equation of the curve determines κ^2 but not κ uniquely. So long as we consider only one radius of curvature, it does not make much difference what sign we attach to κ. The simplest way is to

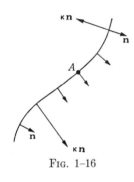

FIG. 1–16

take $\kappa > 0$, that is, to take the sense of the curvature vector as the sense of $\mathbf{n}$. But the sign of κ is of some importance when we consider a family of curvature vectors. For instance, if we take a plane curve (with continuous derivatives) with a point of inflection at A (Fig. 1–16), then the curvature vectors are pointed in different directions on both sides of A and it may be convenient to distinguish the concave and the convex sides of the curve by different signs of κ. The field of unit vectors $\mathbf{n}$ along the curve is then continuous. We shall meet another example in the theory of surfaces where we have many curvature vectors at one point (Sec. 2–6).

When the curve is plane we can remove the ambiguity in the sign of κ by postulating that the sense of rotation from $\mathbf{t}$ to $\mathbf{n}$ is the same as that from OX to OY. Then κ can be defined by the equation $\kappa = d\varphi/ds$, where φ is the angle of the tangent vector with the positive X-axis.

1–5 Torsion.   The curvature measures the rate of change of the tangent when moving along the curve. We shall now introduce a quantity measuring the rate of change of the osculating plane. For this purpose we introduce the normal at P to the osculating plane, the *binormal*. In it we place the *unit binormal vector* $\mathbf{b}$ in such a way that the sense $\mathbf{t} \rightarrow \mathbf{n} \rightarrow \mathbf{b}$ is the same as that of $OX \rightarrow OY \rightarrow OZ$; in other words, since $\mathbf{t}, \mathbf{n}, \mathbf{b}$ are mutually perpendicular unit vectors, we define the vector $\mathbf{b}$ by the formula:

$$\mathbf{b} = \mathbf{t} \times \mathbf{n}. \tag{5–1}$$

These three vectors **t**, **n**, **b** can be taken as a new frame of reference. They satisfy the relations

$$\begin{array}{lll} \mathbf{t} \cdot \mathbf{t} = 1, & \mathbf{n} \cdot \mathbf{n} = 1, & \mathbf{b} \cdot \mathbf{b} = 1, \\ \mathbf{t} \cdot \mathbf{n} = 0, & \mathbf{n} \cdot \mathbf{b} = 0, & \mathbf{b} \cdot \mathbf{t} = 0. \end{array} \tag{5-2}$$

This frame of reference, moving along the curve, forms *the moving trihedron.*

The rate of change of the osculating plane is expressed by the vector

$$\mathbf{b}' = d\mathbf{b}/ds.$$

This vector lies in the direction of the principal normal, since, according to the equation $\mathbf{b} \cdot \mathbf{t} = 0$,

$$\begin{array}{l} \mathbf{b}' \cdot \mathbf{t} + \mathbf{b} \cdot \mathbf{t}' = 0, \\ \mathbf{b}' \cdot \mathbf{t} = -\mathbf{b} \cdot \kappa\mathbf{n} = 0, \end{array}$$

and, because $\mathbf{b} \cdot \mathbf{b} = 1$,

$$\mathbf{b}' \cdot \mathbf{b} = 0,$$

so that, introducing a proportionality factor τ,

$$d\mathbf{b}/ds = -\tau\mathbf{n}. \tag{5-3}$$

We call τ the *torsion* of the curve. It may be positive or negative, like the curvature, but where the equation of the curve defines only κ^2, it does define τ uniquely. This can be shown by expressing τ as follows:

$$\begin{aligned} \tau &= -\mathbf{n} \cdot (\mathbf{t} \times \mathbf{n})' = -\mathbf{n} \cdot (\mathbf{t} \times \mathbf{n}') \\ &= -\kappa^{-1}\mathbf{x}'' \cdot (\mathbf{x}' \times (\kappa^{-1}\mathbf{x}'')') \\ &= \kappa^{-2}(\mathbf{x}'\mathbf{x}''\mathbf{x}'''), \end{aligned}$$

or

$$\tau = \frac{(\mathbf{x}'\mathbf{x}''\mathbf{x}''')}{\mathbf{x}'' \cdot \mathbf{x}''}, \quad \mathbf{x}' = d\mathbf{x}/ds. \tag{5-4}$$

This formula expresses τ in $\mathbf{x}(s)$ and its derivatives independently of the orientation of the curve, since change of s into $-s$ does not affect the right-hand member of (5-4). The sign of τ has therefore a meaning for the nonoriented curve. We shall discuss this further below.

The equations

$$\begin{aligned} \mathbf{x}'(s) &= d\mathbf{x}/ds = (d\mathbf{x}/du)(du/ds) = \dot{\mathbf{x}}u' = \dot{\mathbf{x}}(\dot{\mathbf{x}} \cdot \dot{\mathbf{x}})^{-1/2}, \\ \mathbf{x}'' &= \ddot{\mathbf{x}}(u')^2 + \dot{\mathbf{x}}u'' = [(\dot{\mathbf{x}} \cdot \dot{\mathbf{x}})\ddot{\mathbf{x}} - (\dot{\mathbf{x}} \cdot \ddot{\mathbf{x}})\dot{\mathbf{x}}](\dot{\mathbf{x}} \cdot \dot{\mathbf{x}})^{-2}, \\ \mathbf{x}''' &= \dddot{\mathbf{x}}(u')^3 + 3\ddot{\mathbf{x}}u'u'' + \dot{\mathbf{x}}u''', \end{aligned}$$

allow us to express κ^2 and τ in terms of an arbitrary parameter. We find the formulas

$$\kappa^2 = \frac{(\dot{\mathbf{x}} \times \ddot{\mathbf{x}}) \cdot (\dot{\mathbf{x}} \times \ddot{\mathbf{x}})}{(\dot{\mathbf{x}} \cdot \dot{\mathbf{x}})^3}, \qquad \dot{\mathbf{x}} = d\mathbf{x}/du \quad (5\text{–}5\text{a})$$

$$\tau = \frac{(\dot{\mathbf{x}}\ddot{\mathbf{x}}\dddot{\mathbf{x}})}{(\dot{\mathbf{x}} \times \ddot{\mathbf{x}}) \cdot (\dot{\mathbf{x}} \times \ddot{\mathbf{x}})}. \qquad\qquad (5\text{–}5\text{b})$$

We see that κ and τ have the dimension L^{-1}. Where $|\kappa^{-1}| = |R|$ is called the radius of curvature, $|\tau^{-1}| = |T|$ is called the *radius of torsion*. However, this quantity $|T|$ does not admit of such a ready and elegant geometrical interpretation as $|R|$.

EXAMPLES. (1) *Circular helix.* From Eq. (2–16) we derive, if $\sqrt{a^2+b^2}=c$,

$$\mathbf{x}'\left(-\frac{a}{c}\sin u, \frac{a}{c}\cos u, \frac{b}{c}\right),$$

$$\mathbf{x}''\left(-\frac{a}{c^2}\cos u, -\frac{a}{c^2}\sin u, 0\right), \quad \left(s \text{ parameter}, u = \frac{s}{c}\right)$$

$$\mathbf{x}'''\left(\frac{a}{c^3}\sin u, -\frac{a}{c^3}\cos u, 0\right).$$

Hence

$$\kappa^2 = \mathbf{x}'' \cdot \mathbf{x}'' = a^2/c^4, \quad \kappa = \pm a/c^2,$$

$$\tau = \frac{(\mathbf{x}'\mathbf{x}''\mathbf{x}''')}{\mathbf{x}'' \cdot \mathbf{x}''} = \frac{b}{c} \begin{vmatrix} -\dfrac{a}{c^2}\cos u & -\dfrac{a}{c^2}\sin u \\[2mm] \dfrac{a}{c^3}\sin u & -\dfrac{a}{c^3}\cos u \end{vmatrix} \frac{c^4}{a^2} = \frac{b}{c^2}.$$

Hence τ is positive when b is positive, which is the case when (see Sec. 1–1) the helix is right-handed; τ is negative for a left-handed helix. We also see that κ and τ are both constants, and from the equations

$$a = \frac{\pm \kappa}{\kappa^2 + \tau^2}, \quad b = \frac{\tau}{\kappa^2 + \tau^2}$$

we can derive one and only one circular helix with given κ, τ and with given position with respect to the coordinate axes (change of a into $-a$ does not change the helix; it only changes u into $u + \pi$).

(2) *Plane curve.* Since $\mathbf{b}$ is constant, $\tau = 0$. If, conversely, $\tau = 0$, $(\dot{\mathbf{x}}\ddot{\mathbf{x}}\dddot{\mathbf{x}}) = 0$, or $\dddot{\mathbf{x}} + \lambda\ddot{\mathbf{x}} + \mu\dot{\mathbf{x}} = 0$; λ, μ functions of s. This is a linear homogeneous equation in $\dot{\mathbf{x}}$, which is solved by an expression of the form

$$\dot{\mathbf{x}} = \mathbf{c}_1 f_1(s) + \mathbf{c}_2 f_2(s),^*$$

---

* See e.g. P. Franklin, *Methods of advanced calculus*, New York: McGraw-Hill Book Co., 1941, p. 351. A vector equation is equivalent to three scalar equations, so that the result reached for scalar differential equations can immediately be translated into vector language.

hence

$$\mathbf{x} = \mathbf{c}_1 F_1(s) + \mathbf{c}_2 F_2(s) + \mathbf{c}_3,$$

where the $\mathbf{c}_i$, $i = 1, 2, 3$, are constant vectors and the F_j, $j = 1, 2$, functions determined by the arbitrary λ and μ. This shows that the curve $\mathbf{x}(s)$ lies in the plane through the end point of $\mathbf{c}_3$ parallel to $\mathbf{c}_1$ and $\mathbf{c}_2$. This means that $\mathbf{x}(s)$ can be any plane curve. For straight lines the torsion is indeterminate.

Curvature and torsion are also known as *first* and *second curvature*, and space curves are also known as *curves of double curvature*.

The name *torsion* is due to L. I. Vallée, *Traité de géométrie descriptive*, p. 295 of the 1825 edition. The older term was *flexion*. The name *binormal* is due to B. de Saint Venant, *Journal Ecole Polytechnique* **18**, 1845, p. 17.

1–6 Formulas of Frenet. We have found that $\mathbf{t}' = \kappa\mathbf{n}$ and $\mathbf{b}' = -\tau\mathbf{n}$. Let us complete this information by also expressing $\mathbf{n}' = d\mathbf{n}/ds$ in terms of the unit vectors of the moving trihedron. Since $\mathbf{n}'$ is perpendicular to $\mathbf{n}$, $\mathbf{n} \cdot \mathbf{n}' = 0$, and we can express $\mathbf{n}'$ linearly in terms of $\mathbf{t}$ and $\mathbf{b}$:

$$\mathbf{n}' = \alpha_1\mathbf{t} + \alpha_2\mathbf{b}.$$

Since according to Eq. (5–2)

$$\alpha_1 = \mathbf{t} \cdot \mathbf{n}' = -\mathbf{n} \cdot \mathbf{t}' = -\mathbf{n} \cdot \kappa\mathbf{n} = -\kappa,$$

and

$$\alpha_2 = \mathbf{b} \cdot \mathbf{n}' = -\mathbf{n} \cdot \mathbf{b}' = +\mathbf{n} \cdot \tau\mathbf{n} = \tau,$$

we find for $d\mathbf{n}/ds$:

$$\mathbf{n}' = -\kappa\mathbf{t} + \tau\mathbf{b}.$$

The three vector formulas,

$$\boxed{\begin{aligned}
\frac{d\mathbf{t}}{ds} &= \qquad\quad \kappa\mathbf{n} \\
\frac{d\mathbf{n}}{ds} &= -\kappa\mathbf{t} \qquad\quad +\tau\mathbf{b} \\
\frac{d\mathbf{b}}{ds} &= \qquad\quad -\tau\mathbf{n}
\end{aligned}}, \tag{6–1}$$

together with $d\mathbf{x}/ds = \mathbf{t}$, describe the motion of the moving trihedron along the curve. They take a central position in the theory of space curves and are known as the *formulas of Frenet*, or of *Serret-Frenet*.

They were obtained in the Toulouse dissertation of F. Frenet, 1847, of which an abstract appeared as "Sur les courbes à double courbure," *Journal de Mathém.* **17** (1852), pp. 437–447. The paper of J. A. Serret appeared *Journal de Mathém.* **16** (1851), pp. 193–207; it appeared after Frenet's thesis, but before Frenet made his results more widely known.

The coordinates of **t**, **n**, and **b** are the cosines of the angles which the oriented tangent, principal normal, and binormal make with the positive coordinate axes. When we indicate this by $\mathbf{t}(\cos \alpha_i)$, $\mathbf{n}(\cos \beta_i)$, $\mathbf{b}(\cos \gamma_i)$, $i = 1, 2, 3$, the Frenet formulas take the following coordinate form:

$$\frac{d}{ds} \cos \alpha_i = \kappa \cos \beta_i, \qquad \frac{d}{ds} \cos \beta_i = -\kappa \cos \alpha_i + \tau \cos \gamma_i,$$

$$\frac{d}{ds} \cos \gamma_i = -\tau \cos \beta_i. \tag{6-1a}$$

The three planes formed by the three sides of the moving trihedron (Fig. 1–17) are called:

the *osculating plane*, through tangent and principal normal, with equation

$$(\mathbf{y} - \mathbf{x}) \cdot \mathbf{b} = 0,$$

the *normal plane*, through principal normal and binormal, with equation

$$(\mathbf{y} - \mathbf{x}) \cdot \mathbf{t} = 0,$$

the *rectifying plane*, through binormal and tangent, with equation

$$(\mathbf{y} - \mathbf{x}) \cdot \mathbf{n} = 0.$$

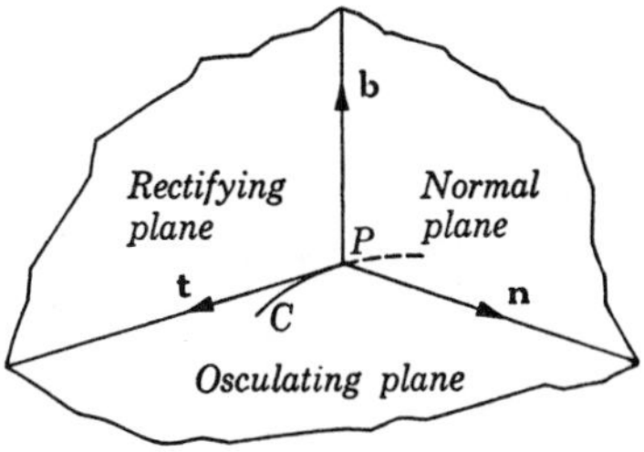

FIG. 1–17

If we take the moving trihedron at P as the trihedron of a set of new cartesian coordinates x, y, z, then the behavior of the curve near P is expressed by the formulas (6–1a) in the form ($\mathbf{x}'' = \kappa \mathbf{n}$, $\mathbf{x}''' = -\kappa^2 \mathbf{t} + \kappa' \mathbf{n} + \kappa\tau\mathbf{b}$):

$$\begin{aligned}
x' &= 1, & y' &= 0, & z' &= 0, \\
x'' &= 0, & y'' &= \kappa, & z'' &= 0, \\
x''' &= -\kappa^2, & y''' &= \kappa', & z''' &= \kappa\tau.
\end{aligned} \tag{6-2}$$

From these equations we deduce for $s \to 0$

$$\lim \frac{y}{x^2} = \lim \frac{y'}{2xx'} = \lim \frac{y''}{2x'^2} = \frac{\kappa}{2}, \tag{6-3}$$

$$\lim \frac{z}{x^3} = \lim \frac{z'}{3x^2x'} = \lim \frac{z''}{6x(x')^2} = \lim \frac{z'''}{6(x')^3} = \frac{\kappa\tau}{6};$$

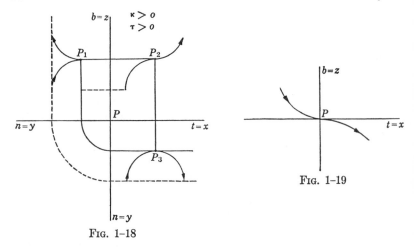

Fig. 1–18

Fig. 1–19

hence

$$\lim \frac{z^2}{y^3} = \frac{\kappa^2 \tau^2}{36} \frac{8}{\kappa^3} = \frac{2}{9} \tau^2 \kappa^{-1} = \frac{2}{9} R \tau^2.$$

This shows that the projections of the curve on the three planes of the moving trihedron behave near P like the curves

$$y = \frac{\kappa}{2} x^2 \quad \text{(projection on the osculating plane)},$$

$$z = \frac{\kappa \tau}{6} x^3 \quad \text{(projection on the rectifying plane)}, \quad (6\text{–}4)$$

$$z^2 = \frac{2}{9} \tau^2 R y^3 \quad \text{(projection on the normal plane)}.$$

Fig. 1–18 shows this behavior in an orthographic projection, taking $\kappa > 0$, $\tau > 0$. If the sign of τ is changed, the projection on the rectifying plane changes to that of Fig. 1–19. This again shows the geometrical meaning of the sign of τ.

Fig. 1–20 gives a representation of the curve and its trihedron in space.

EXERCISES

1. Find the curvature and the torsion of the curves:

(a) $$x = u, \quad y = u^2, \quad z = u^3.$$

(b) $$x = u, \quad y = \frac{1 + u}{u}, \quad z = \frac{1 - u^2}{u}. \quad \text{(Why is } \tau = 0?)$$

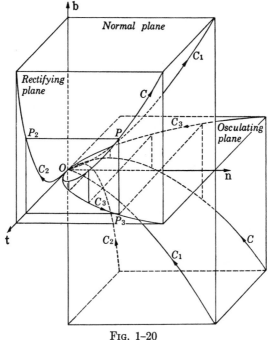

F<small>IG</small>. 1–20

(c) $$y = f(x), \qquad z = g(x).$$
(d) $$x = a(u - \sin u), \qquad y = a(1 - \cos u), \qquad z = bu.$$
(e) $$x = a(3u - u^3), \qquad y = 3au^2, \qquad z = a(3u + u^3) \quad \text{(Here } \kappa^2 = \tau^2.\text{)}$$

2. (a) When all the tangent lines of a curve pass through a fixed point, the curve is a straight line.   (b) When they are parallel to a given line the curve is also a straight line.

3. (a) When all the osculating planes of a curve pass through a fixed point, the curve is plane.   (b) When they are parallel to a given plane the curve is also plane.

4. The binormal of a circular helix makes a constant angle with the axis of the cylinder on which the helix lies.

5. Show that the tangents to a space curve and to the locus C of its centers of curvature at corresponding points are normal.

6. The locus C of the centers of curvature of a circular helix is a coaxial helix of equal pitch.

7. Show that the locus of the centers of curvature of the locus C of problem 6 is the original circular helix and that the product of the torsions at corresponding points of C and the helix is equal to κ^2, the square of the curvature of the helix.

8. All osculating planes to a circular helix passing through a given point not lying on the helix have their points of contact with the helix in the same plane.

9. Prove that the curve

$$x = a \sin^2 u, \qquad y = a \sin u \cos u, \qquad z = a \cos u$$

lies on a sphere, and verify that all normal planes pass through the origin. Show that this curve is of degree four.

10. When $\mathbf{x} = \mathbf{x}(t)$ is the path of a moving point as a function of time, show that the acceleration vector lies in the osculating plane.

11. Determine the condition that the osculating circle passes through four consecutive points of the curve.

12. Show that for a plane curve, for which $x = x(s)$, $y = y(s)$, and the sign of κ is determined by the assumption of p. 15, $\kappa = x'y'' - x''y'$.

13. Starting with the Eq. (1–5) for the curve $\mathbf{x}(s)$, derive expansions for the projections on the three planes of the moving trihedron, and compare with Eq. (6–4).

14. Determine the form of the function $\varphi(u)$ such that the principal normals of the curve $x = a \cos u, y = a \sin u, z = \varphi(u)$ are parallel to the XOY-plane.

15. (a) The binormal at a point P of the curve is the limiting position of the common perpendicular to the tangents at P and a neighboring point Q, when $Q \to P$. Also find (b) the limiting position for the common perpendicular to the binormals.

16. Find the unit tangent vector of the curve given by

$$F_1(x, y, z) = 0, \qquad F_2(x, y, z) = 0.$$

17. *Transformation of Combescure.* Two space curves are said to be obtainable from each other by such a transformation if there exists a one-to-one correspondence between their points so that the osculating planes at corresponding points are parallel. Show that the tangents, principal normals, and binormals are parallel. (Following L. Bianchi, *Lezioni I*, p. 50, we call such transformations after E. J. C. Combescure, *Annales École Normale* **4**, 1867, though the transformations discussed in this paper are more specifically qualified, and deal with certain triply orthogonal systems of curves.)

18. *Cinematical interpretation of Frenet's formulas.* When a rigid body rotates about a point there exists an axis of instantaneous rotation, that is, the locus of the points which stay in place. Show that this axis for the moving trihedron (we do not consider the translation expressed by $d\mathbf{x} = \mathbf{t} \, ds$) has the direction of the vector $\mathbf{R} = \tau\mathbf{t} + \kappa\mathbf{b}$, so that the Frenet formulas can be written in the form

$$\mathbf{t}' = \mathbf{R} \times \mathbf{t}, \qquad \mathbf{n}' = \mathbf{R} \times \mathbf{n}, \qquad \mathbf{b}' = \mathbf{R} \times \mathbf{b}.$$

This constitutes the approach to the theory of curves (and surfaces) typical of G. Darboux and E. Cartan, the "méthode du trièdre mobile." (Compare G. Darboux, *Leçons* I; E. Cartan, *La méthode du répère mobile, Actualités scientifiques* **194**, Paris, 1935.)

19. *Spherical image.* When we move all unit tangent vectors **t** of a curve C to a point, their end points will describe a curve on the unit sphere, called the *spherical image* (spherical indicatrix) of C. Show that the absolute value of the curvature is equal to the ratio of the arc length ds_t of the spherical image and the arc length of the curve ds. What is the spherical image (a) of a straight line; (b) of a plane curve; (c) of a circular helix?

20. *Third curvature.* When we extend the operation of Exercise 19 to the vectors **n** and **b**, we obtain the spherical image of the principal normals and of the binormals. If ds_n and ds_b represent the elements of arc of these two images respectively, show that $|ds_n/ds| = +\sqrt{\kappa^2 + \tau^2}$ and $|ds_b/ds| = |\tau|$. The quantity $\sqrt{\kappa^2 + \tau^2}$ is sometimes called the *third* (or *total*) *curvature.*

1–7 Contact. Instead of stating that figures have a certain number of consecutive points (or other elements) in common, we can also state that they have a *contact* of a certain order. The general definition is as follows (Fig. 1–21):

Let two curves or surfaces Σ_1, Σ_2 have a regular point P in common. Take a point A on Σ_1 near P and let AD be its distance to Σ_2. Then Σ_2 has a contact of order n with Σ_1 at P, when for $A \to P$ along Σ_1

$$\lim \frac{AD}{(AP)^k}$$

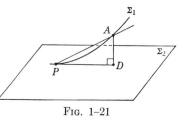

Fig. 1–21

is finite $(\neq 0)$ *for* $k = n + 1$, *but* $= 0$ *for* $k = n$. $[AD = \mathrm{o}((AP)^k)$ *for* $k = 1, 2, \ldots, n]$

When Σ_1 is a curve $\mathbf{x}(u)$ and Σ_2 a surface $(F_x, F_y, F_z$ not all zero$)$

$$F(x, y, z) = 0, \qquad (7\text{–}1)$$

we make use of the fact that the distance AD of a point $A(x_1, y_1, z_1)$ near P is of the same order as $F(x_1, y_1, z_1)$. The general proof of this fact requires some surface theory, but in the case of the plane and the sphere, the only cases we discuss in the text, it can be readily demonstrated (see Exercise 4, Sec. 1–11).

Let us now consider the function obtained by substituting the x_i of the curve Σ_1 into Eq. (7–1):

$$f(u) = F[x(u), y(u), z(u)]. \qquad (7\text{–}2)$$

This procedure is simply a generalization of the method used in Secs. 1–3 and 1–4 to obtain the equations of the osculating plane and the osculating circle. Let $f(u)$ near $P(u = u_0)$ have finite derivatives $f^{(i)}(u_0)$, $i = 1, 2,$

$\ldots, n + 1$. Then if we take $u = u_1$ at A and write $h = u_1 - u_0$, then there exists a Taylor development of $f(u)$ of the form (compare Eq. (1–5)):

$$f(u_1) = f(u_0) + hf'(u_0) + \frac{h^2}{2!}f''(u_0) + \cdots + \frac{h^{n+1}}{(n+1)!}f^{(n+1)}(u_0) + o(h^{n+1}).$$
$$(7\text{–}3)$$

Here $f(u_0) = 0$, since P lies on Σ_2, and h is of order AP (see theorem Sec. 1–2); $f(u_1)$ is of the order of AD. *Hence necessary and sufficient conditions that the surface has a contact of order n at P with the curve are that at P the relations hold:*

$$f(u) = f'(u) = f''(u) = \cdots = f^{(n)}(u) = 0; \quad f^{(n+1)}(u) \neq 0. \quad (7\text{–}4)$$

In these formulas we have replaced u_0 by u.

In the same way we find, if Σ_2 is a curve defined by

$$F_1(x, y, z) = 0, \qquad F_2(x, y, z) = 0,$$

that *necessary and sufficient conditions for a contact of order n at P between the curves are that at P*

$$f_1(u) = f_1'(u) = \cdots = f_1^{(n)}(u) = 0, \qquad (7\text{–}5)$$
$$f_2(u) = f_2'(u) = \cdots = f_2^{(n)}(u) = 0,$$

where

$$f_1(u) = F_1[x(u), y(u), z(u)], \qquad f_2(u) = F_2[x(u), y(u), z(u)]$$

and at least one of the two derivatives $f_1^{(n+1)}(u), f_2^{(n+1)}(u)$ at P does not vanish. We can develop similar conditions for the contact of two surfaces.

Instead of AD we can use segments of the same order, making, for instance, $PD = PA$ (then $\angle D$ no longer $90°$).

If we compare these conditions with our derivation of osculating plane and osculating circle, then we see that they are, in these cases, identical with the condition that Σ_1 and Σ_2 have $n + 1$ consecutive points in common. And so we can say that, in general:

Two figures Σ_1 and Σ_2, having at P a contact of order n, have $n + 1$ consecutive points in common.

Indeed, following again a reasoning analogous to that of Secs. 1–3 and 1–4, and confining ourselves to the case expressed by Eq. (7–1), let us define F with $n + 1$ independent parameters. These are just enough to let surface Σ_2 pass through $(n + 1)$ points $(u_0, u_1, \ldots, u_n)$. If these $n + 1$ points come together in point $u = u_0$, then the $n + 1$ equations (7–4) are satisfied; if there were more such equations (7–4), then we could determine the parameters in F so that Σ_2 would pass through more than $n + 1$ points.

A similar reasoning holds for the other cases of contact. From this and the theorems of Secs. 1–3 and 1–4 follows:

A tangent has a contact of (at least) order one with the curve.

An osculating plane and an osculating circle have a contact of (at least) order two with the curve.

The study of the contact of curves and surfaces was undertaken in considerable detail in Lagrange's *Traité des fonctions analytiques* (1797) and in Cauchy's *Leçons sur les applications du calcul infinitésimal à la géométrie I* (1826).

We shall now apply this theory to find a sphere passing through four consecutive points of the curve, the *osculating sphere*. Let this sphere be given by the equation

$$(\mathbf{X} - \mathbf{c}) \cdot (\mathbf{X} - \mathbf{c}) - r^2 = 0, \quad (\mathbf{X} \text{ generic point of the sphere}, \mathbf{c} \text{ its center}, r \text{ radius}).$$

Consider, in accordance with Eq. (7–2):

$$f(s) = (\mathbf{x} - \mathbf{c}) \cdot (\mathbf{x} - \mathbf{c}) - r^2, \quad \mathbf{x} = \mathbf{x}(s).$$

Then the Eqs. (7–4) take the form, apart from $f(s) = 0$:

$$f'(s) = 0, \quad \text{or} \quad (\mathbf{x} - \mathbf{c}) \cdot \mathbf{t} = 0,$$
$$f''(s) = 0, \quad \text{or} \quad (\mathbf{x} - \mathbf{c}) \cdot \kappa \mathbf{n} + 1 = 0,$$
$$f'''(s) = 0, \quad \text{or} \quad (\mathbf{x} - \mathbf{c}) \cdot (\kappa' \mathbf{n} - \kappa^2 \mathbf{t} + \kappa \tau \mathbf{b}) = 0,$$

or ($\tau \neq 0$)

$$(\mathbf{x} - \mathbf{c}) \cdot \mathbf{t} = 0, \quad (\mathbf{x} - \mathbf{c}) \cdot \mathbf{n} = -R, \quad (\mathbf{x} - \mathbf{c}) \cdot \mathbf{b} = R\kappa'/\kappa\tau = -R'T.$$

The center O of the osculating sphere is thus uniquely determined by

$$\mathbf{c} = \mathbf{x} + R\mathbf{n} + TR'\mathbf{b}. \tag{7-6}$$

This sphere has a contact of order three with the curve. Its intersection with the osculating plane is the osculating circle. Its center lies in the normal plane (Fig. 1–22) on a line parallel to the binormal, the *polar axis*. The radius of the osculating sphere is

$$r = \sqrt{R^2 + (TR')^2}. \tag{7-7}$$

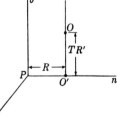

When the curve is of constant curvature (not a circle), the center of the osculating sphere coincides with the center of the osculating circle.

The result expressed by Eq. (7–6) is due to Monge (1807), see his *Applications* (1850), p. 412. Monge's notation is quite different.

Fig. 1–22

1–8 Natural equations. When a curve is defined by an equation $\mathbf{x} = \mathbf{x}(s)$, its form depends on the choice of the coordinate system. When a curve is moved without change in its shape, its equation with respect to the coordinate system changes. It is not always immediately obvious whether two equations represent the same curve except for its position with respect to the coordinate system. Even in the simple case of equations of the second degree in the plane (conics) such a determination requires some work. The question therefore arises: Is it possible to characterize a curve by a relation independent of the coordinates? This can actually be accomplished; such an equation is called *natural* or *intrinsic*.

It is easily seen that a relation between curvature and arc length gives a natural equation for a plane curve. Indeed, if we give an equation of the form

$$\kappa = \kappa(s), \tag{8–1}$$

then we find, using the relations

$$R^{-1} = \kappa = d\varphi/ds, \qquad \cos \varphi = dx/ds, \qquad \sin \varphi = dy/ds,$$

that x and y can be found by two quadratures:

$$x = \int_{\varphi_0}^{\varphi} R \cos \varphi \, d\varphi, \qquad y = \int_{\varphi_0}^{\varphi} R \sin \varphi \, d\varphi, \qquad \varphi = \int_{s_0}^{s} \kappa \, ds. \tag{8–2}$$

Change of integration constant in x and y means translation, change of integration constant in φ means rotation of the curve, and thus we can obtain all possible equations in rectangular coordinates, selecting in each case the most convenient one for our purpose.

This representation of a curve by means of κ (or R) and s goes back to Euler, who used it for special curves: *Comment. Acad. Petropolit.* **8**, 1736, pp. 66–85. The choice of κ and s as natural coordinates can be criticized, since s still contains an arbitrary constant and κ is determined but for the sign. G. Scheffers has therefore developed a system of natural equations of a plane curve in which $d(\kappa^2)/d\varphi$ is expressed as a function of κ^2. See *Anwendung I*, pp. 84–91.

Examples. (1) *Circle:* $\kappa = a = \text{const.}$

$$x = R \sin \varphi, \qquad y = -R \cos \varphi, \qquad \text{if } u = \varphi - \frac{\pi}{2},$$
$$x = R \cos u, \qquad y = R \sin u. \qquad \text{(Fig. 1–23)}$$

When $a = 0$ we obtain a straight line.

(2) *Logarithmic spiral.*

$$R = as + b = s \cot \alpha + p \csc \alpha \quad (\alpha, p \text{ constants}),$$

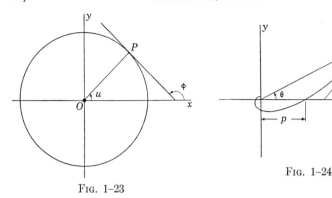

Fig. 1–24

Fig. 1–23

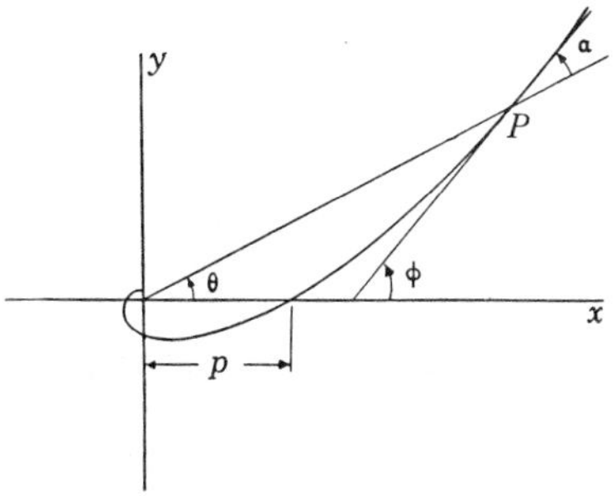

$$\varphi = \alpha + (\tan \alpha) \ln \frac{s \cos \alpha + p}{p}, \quad \text{(selecting } s = 0 \text{ for } \varphi = \alpha),$$

$$R = p(\csc \alpha) \exp (\theta \cot \alpha), \qquad \theta = \varphi - \alpha. \quad \text{(Fig. 1–24)}$$

Introducing polar coordinates $x + iy = re^{i\theta}$, we find

$$r = p \exp (\theta \cot \alpha) \quad (x = p, y = 0 \text{ for } s = 0, \varphi = \alpha).$$

(3) *Circle involute:*

$$R^2 = 2as; \qquad \varphi = \sqrt{2/a}\ \sqrt{s}, \qquad s = \tfrac{1}{2}a\varphi^2,$$

$$R = a\varphi, \qquad x = a \int \varphi \cos \varphi \, d\varphi = a(\cos \varphi + \varphi \sin \varphi),$$

$$y = a \int \varphi \sin \varphi \, d\varphi = a(\sin \varphi - \varphi \cos \varphi). \quad \text{(Fig. 1–25)}$$

(4) *Epicycloid.* We start with the equations in x and y:

$$x = (a + b) \cos \psi - b \cos \frac{a + b}{b} \psi,$$

$$y = (a + b) \sin \psi - b \sin \frac{a + b}{b} \psi,$$

$$s = 4 \frac{b(a + b)}{a} \cos \frac{a\psi}{2b},$$

$$R = \frac{4b(a + b)}{a + 2b} \sin \frac{a\psi}{2b}.$$

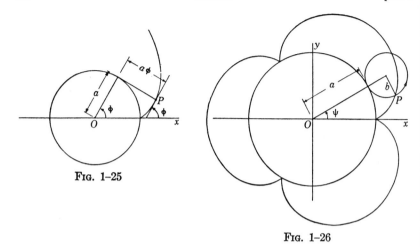

Fig. 1–25

Fig. 1–26

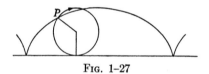

Fig. 1–27

Hence the natural equation is:

$$\frac{s^2}{A^2} + \frac{R^2}{B^2} = 1, \tag{8–3}$$

where

$$A = \frac{4b(a+b)}{a}, \quad B = \frac{4b(a+b)}{a+2b}; \quad A > B. \quad \text{(Fig. 1–26)} \tag{8–4}$$

When $A < B$ we obtain a hypocycloid. An epicycloid is the locus of a point on the circumference of a circle when rolling on a fixed circle on the outside; a hypocycloid when rolling on the inside. When A/B is rational, the curve is closed. When $a \to \infty$, $A \to 4b$, $B \to 4b$, and we obtain

$$s^2 + R^2 = 16b^2,$$

the equation of a *cycloid*, obtained by rolling a circle of radius b on a straight line (Fig. 1–27). We now prove the

FUNDAMENTAL THEOREM for space curves:

If two single-valued continuous functions $\kappa(s)$ and $\tau(s)$, $s > 0$, are given, then there exists one and only one space curve, determined but for its position in space, for which s is the arc length (measured from an appropriate point on the curve), κ the curvature, and τ the torsion.

The equations $\kappa = \kappa(s)$, $\tau = \tau(s)$ are the *natural* or *intrinsic* equations of the space curve.

The proof is simple when we confine ourselves to *analytic* functions. Then we can write, in the neighborhood of a point $s = s_0$, $h = s - s_0$:

$$\mathbf{x}(s) = \mathbf{x}(s_0) + \frac{h}{1}\mathbf{x}'(s_0) + \frac{h^2}{2!}\mathbf{x}''(s_0) + \cdots,$$

provided the series is convergent in a certain interval $s_1 < s_0 < s_2$. Then, substituting for $\mathbf{x}'$, $\mathbf{x}''$, etc., their values with respect to the moving trihedron at $P(s_0)$, we obtain

$$\mathbf{x}' = \mathbf{t}, \qquad \mathbf{x}'' = \kappa\mathbf{n}, \qquad \mathbf{x}''' = -\kappa^2\mathbf{t} + \kappa'\mathbf{n} + \kappa\tau\mathbf{b}, \qquad \mathbf{x}'''' = \cdots,$$

so that

$$\mathbf{x}(s) = \mathbf{x}(s_0) + h\mathbf{t} + \tfrac{1}{2}\kappa h^2\mathbf{n} + \tfrac{1}{6}h^3(-\kappa^2\mathbf{t} + \kappa'\mathbf{n} + \kappa\tau\mathbf{b}) + \cdots, \qquad (8\text{--}5)$$

where all terms can be successively found by differentiating the Frenet formulas, and all successive derivatives of κ and τ taken, as well as $\mathbf{t}$, $\mathbf{n}$, $\mathbf{b}$, at $P(s_0)$ are supposed to exist because of the analytical character of the functions. If we now choose at an arbitrary point $\mathbf{x}(s_0)$ an arbitrary set of three mutually perpendicular unit vectors and select them as $\mathbf{t}$, $\mathbf{n}$, $\mathbf{b}$, then Eq. (8–5) determines the curve uniquely (inside the interval of convergence).

It is, however, possible to prove the theorem under the sole assumption that $\kappa(s)$ and $\tau(s)$ are continuous. In this case we apply to the system of three simultaneous differential equations of the first order in α, β, γ,

$$\frac{d\alpha}{ds} = \kappa\beta, \qquad \frac{d\beta}{ds} = -\kappa\alpha + \tau\gamma, \qquad \frac{d\gamma}{ds} = -\tau\beta, \qquad (8\text{--}6)$$

the theorem concerning the existence of solutions.

This theorem is as follows. Given a system of differential equations

$$\frac{dy_i}{dx} = f_i(x, y_1, \cdots y_n), \qquad i = 1, 2, \cdots n,$$

where the f_i are single-valued and continuous in their $n + 1$ arguments inside a given interval (with a Lipschitz condition, satisfied in our case). Then there exists a unique set of continuous solutions of this system which assumes given values $y_1^0, y_2^0, \ldots y_n^0$ when $x = x_0$.*

We deduce from this theorem that we can find in one and in only one way three continuous solutions $\alpha_1(s), \beta_1(s), \gamma_1(s)$ which assume for $s = s_0$ the values $1, 0, 0$ respectively. We can similarly find three continuous solutions $\alpha_2, \beta_2, \gamma_2$, so that

$$\alpha_2(s_0) = 0, \qquad \beta_2(s_0) = 1, \qquad \gamma_2(s_0) = 0,$$

and three more continuous solutions $\alpha_3, \beta_3, \gamma_3$, so that

$$\alpha_3(s_0) = 0, \qquad \beta_3(s_0) = 0, \qquad \gamma_3(s_0) = 1.$$

The Eqs. (8–6) lead to the following relations between the α, β, γ:

$$\frac{1}{2}\frac{d}{ds}(\alpha_1^2 + \beta_1^2 + \gamma_1^2) = \kappa\beta_1\alpha_1 - \kappa\alpha_1\beta_1 + \tau\gamma_1\beta_1 - \tau\beta_1\gamma_1 = 0,$$

or

$$\alpha_1^2 + \beta_1^2 + \gamma_1^2 = \text{const} = 1 + 0 + 0 = 1. \tag{8–7a}$$

Similarly, we find two more relations of the same form:

$$\alpha_2^2 + \beta_2^2 + \gamma_2^2 = 1, \qquad \alpha_3^2 + \beta_3^2 + \gamma_3^2 = 1, \tag{8–7b}$$

and the three additional relations:

$$\begin{aligned}
\alpha_1\alpha_2 + \beta_1\beta_2 + \gamma_1\gamma_2 &= 0,\\
\alpha_1\alpha_3 + \beta_1\beta_3 + \gamma_1\gamma_3 &= 0,\\
\alpha_2\alpha_3 + \beta_2\beta_3 + \gamma_2\gamma_3 &= 0.
\end{aligned} \tag{8–8}$$

We have thus found a set of mutually perpendicular unit vectors

$$\mathbf{t}(\alpha_1\alpha_2\alpha_3), \qquad \mathbf{n}(\beta_1\beta_2\beta_3), \qquad \mathbf{b}(\gamma_1\gamma_2\gamma_3),$$

where the $\alpha_i, \beta_i, \gamma_i$ all are functions of the parameter s $(i = 1, 2, 3)$.

This is the consequence of the theorem that if the relations (8–7) and (8–8) hold, the relations

$$\Sigma\alpha_i^2 = \Sigma\beta_i^2 = \Sigma\gamma_i^2 = 1, \qquad \Sigma\alpha_i\beta_i = \Sigma\alpha_i\gamma_i = \Sigma\beta_i\gamma_i = 0$$

also hold. This means geometrically that, when $\mathbf{t}, \mathbf{n}, \mathbf{b}$ are three mutually orthogonal unit vectors defined with reference to the set $\mathbf{e}_1, \mathbf{e}_2, \mathbf{e}_3$, then $\mathbf{e}_1, \mathbf{e}_2, \mathbf{e}_3$ are three mutually orthogonal unit vectors defined with reference to the set $\mathbf{t}, \mathbf{n}, \mathbf{b}$.

* Compare E. L. Ince, *Ordinary differential equations*, Longmans, Green and Co., London, 1927, p. 71.

There are ∞^1 trihedrons $(\mathbf{t}, \mathbf{b}, \mathbf{n})$. If we now integrate $\mathbf{t}$, then the equation

$$\mathbf{x} = \int_{s_0}^{s} \mathbf{t} \, ds \tag{8–9}$$

determines a curve which has not only $\mathbf{t}$ as unit tangent vector field, but because of Eq. (8–6) also has $(\mathbf{t}, \mathbf{n}, \mathbf{b})$ as its moving trihedron, κ and τ being its curvature and torsion, and s, because of Eq. (8–9), its arc length. Hence there exists one curve C with given $\kappa(s)$ and $\tau(s)$ of which the moving trihedron at $P(s_0)$ coincides with the coordinate axes.

We now must show that every other curve $\bar{C}$ which can be brought into a one-to-one correspondence with C such that at corresponding points, given by equal s, the curvature and torsion are equal, is congruent to C. This means that $\bar{C}$ can be made to coincide with C by a motion in space. Let us move the point $s = 0$ of $\bar{C}$ to the point $s = 0$ of C (the origin) in such a way that the trihedron $(\mathbf{t}, \mathbf{n}, \mathbf{b})$ of $\bar{C}$ coincides with the trihedron $(\mathbf{t}, \mathbf{n}, \mathbf{b})$ of C (the system $\mathbf{e}_1, \mathbf{e}_2, \mathbf{e}_3$). Let $(\bar{x}_i, \bar{\alpha}_i, \bar{\beta}_i, \bar{\gamma}_i)$ and $(x_i, \alpha_i, \beta_i, \gamma_i)$ now denote the corresponding elements of the moving trihedron of $\bar{C}$ and C respectively. Then Eq. (8–6) holds for $(\bar{\alpha}_i, \bar{\beta}_i, \bar{\gamma}_i)$ and for $(\alpha_i, \beta_i, \gamma_i)$ with the same $\kappa(s)$ and $\tau(s)$. Hence (we omit the index i for a moment):

$$\bar{\alpha}\frac{d\alpha}{ds} + \alpha\frac{d\bar{\alpha}}{ds} + \bar{\beta}\frac{d\beta}{ds} + \beta\frac{d\bar{\beta}}{ds} + \bar{\gamma}\frac{d\gamma}{ds} + \gamma\frac{d\bar{\gamma}}{ds} = \kappa\bar{\alpha}\beta + \kappa\alpha\bar{\beta} - \kappa\bar{\beta}\alpha + \tau\bar{\beta}\gamma$$
$$- \kappa\beta\bar{\alpha} + \tau\beta\bar{\gamma} - \tau\bar{\gamma}\beta - \tau\gamma\bar{\beta} = 0,$$

or

$$\alpha\bar{\alpha} + \beta\bar{\beta} + \gamma\bar{\gamma} = \text{const.}$$

This constant is 1, since it is 1 for $s = 0$. For the $\alpha_i, \ldots, \bar{\gamma}_i$ the equations hold:

$$\alpha_i\bar{\alpha}_i + \beta_i\bar{\beta}_i + \gamma_i\bar{\gamma}_i = 1, \qquad \alpha_i^2 + \beta_i^2 + \gamma_i^2 = 1, \qquad \bar{\alpha}_i^2 + \bar{\beta}_i^2 + \bar{\gamma}_i^2 = 1,$$

which is equivalent to the statement that the three vector pairs $(\alpha_i, \beta_i, \gamma_i)$, $(\bar{\alpha}_i, \bar{\beta}_i, \bar{\gamma}_i)$ make the angle zero with one another. Hence $\alpha_i = \bar{\alpha}_i, \beta_i = \bar{\beta}_i$, $\gamma_i = \bar{\gamma}_i$ for all values of s, so that

$$\frac{d}{ds}(\bar{x}_i - x_i) = 0.$$

This shows that $\bar{x}_i - x_i = \text{const}$, but this constant is zero, since it is zero for $s = 0$. The curves $\bar{C}$ and C coincide, so that the proof of the fundamental theorem is completed.

All curves of given $\kappa(s)$ and $\tau(s)$ can thus be obtained from each other by a motion of space. The resulting curves are at least three times differentiable.

EXAMPLES. (1) *Plane curve.* Here $\kappa(s)$ may be any function, $\tau = 0$. In particular, if κ is constant, we find from Eq. (8–2) the *circle.*

(2) *Circular helix.* $\kappa = $ const, $\tau = $ const. We see this immediately from Example 1, Sec. (1–5), since

$$a = \frac{\kappa}{\kappa^2 + \tau^2} \quad \text{and} \quad b = \frac{\tau}{\kappa^2 + \tau^2}$$

uniquely determine the curve

$$x = a \cos s/c, \qquad y = a \sin s/c, \qquad z = bs/c, \qquad c = \sqrt{a^2 + b^2},$$

and all other curves of the same given κ and τ must be congruent to this curve.

Another way to show this is indicated in Sec. 1–9.

(3) *Spherical curves* (curves lying on a sphere). These are all curves which satisfy the differential equation in natural coordinates

$$R^2 + (TR')^2 = a^2, \qquad a = \text{any constant.} \qquad (8–10)$$

Indeed, when a curve is spherical, its osculating spheres all coincide with the sphere on which the curve lies, hence Eq. (8–10) holds, where a is the radius of the sphere (according to Eq. (7–7)). Conversely, if Eq. (8–10) holds, then the radius of the osculating sphere is constant. Moreover, according to Eq. (7–6):

$$\mathbf{c}' = \mathbf{t} + (-\mathbf{t} + R\tau\mathbf{b}) + R'\mathbf{n} + (TR')'\mathbf{b} - R'\mathbf{n} = \{R\tau + (TR')'\}\mathbf{b} = 0. \quad (8–11)$$

Differentiation of Eq. (8–10) shows that for $\tau \neq 0$, $R' \neq 0$,

$$R\tau + (TR')' = 0, \qquad (8–12)$$

so that $\mathbf{c}' = 0$, and this means that the center of the osculating sphere remains in place (except for $\tau = 0$, $R' = 0$, the circle). Eq. (8–12) is therefore the differential equation of all spherical curves. The circle fits in for $\tau = 0$, $R' = 0$, provided $TR' = 0$.

From Eq. (8–11) it follows, incidentally, that for a nonspherical curve (not plane) the tangent to the locus of the centers of the osculating spheres has the direction of the binormal.

For more information on natural equations see E. Cesaro, *Lezioni di geometria intrinseca,* Napoli, 1896; translated by G. Kowalewski under the title: *Vorlesungen über natürliche Geometrie,* Leipzig, 1901, 341 pp. See also L. Braude, *Les coordonnées intrinsèques,* collection "Scientia," Paris, 1914, 100 pp. Other forms of the natural equations of a space curve can be found in G. Scheffers, *Anwendung I*, pp. 278–287, where the fundamental theorem is proved for $(d\kappa/ds)^2 = f(\kappa^2)$, $\tau = f(\kappa^2)$.

1-9 Helices. The circular helix is a special case of a larger class of curves called (*cylindrical*) *helices* or *curves of constant slope* (German: *Böschungslinien*). They are defined by the property that the tangent makes a constant angle α with a fixed line l in space (the *axis*). Let a unit vector **a** be placed in the direction of l (Fig. 1–28). Then a helix is defined by

$$\mathbf{t} \cdot \mathbf{a} = \cos \alpha = \text{const.}$$

Hence, using the Frenet formulas:

$$\mathbf{a} \cdot \mathbf{n} = 0.$$

a is therefore parallel to the rectifying plane of the curve, and can be written in the form (Fig. 1–29):

$$\mathbf{a} = \mathbf{t} \cos \alpha + \mathbf{b} \sin \alpha,$$

which, differentiated, gives

$$0 = \kappa\mathbf{n} \cos \alpha - \tau\mathbf{n} \sin \alpha = (\kappa \cos \alpha - \tau \sin \alpha)\mathbf{n},$$

or

$$\kappa/\tau = \tan \alpha, \text{ constant.}$$

For curves of constant slope the ratio of curvature to torsion is constant. Conversely, if for a regular curve this condition is satisfied, then we can always find a constant angle α such that

$$\mathbf{n}(\kappa \cos \alpha - \tau \sin \alpha) = 0,$$

$$\frac{d}{ds} (\mathbf{t} \cos \alpha + \mathbf{b} \sin \alpha) = 0,$$

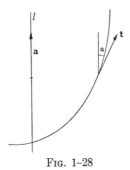

FIG. 1–28   FIG. 1–29

or

$$\mathbf{t} \cos \alpha + \mathbf{b} \sin \alpha = \mathbf{a}, \quad \text{constant unit vector, along the axis.}$$

Hence:

$$\cos \alpha = \mathbf{a} \cdot \mathbf{t}.$$

The curve is therefore of constant slope. We can express this result as follows:

A necessary and sufficient condition that a curve be of constant slope is that the ratio of curvature to torsion be constant. (Theorem of Lancret, 1802; first proof by B. de Saint Venant, *Journal Ec. Polyt.* **30,** 1845, p. 26.)

The equation of a helix can be written in the form (line l is here the Z-axis):

$$x = x(s), \qquad y = y(s), \qquad z = s \cos \alpha,$$

which shows that this curve can be considered as a curve on a general cylinder making a constant angle with the generating lines (*cylindrical loxodrome*). When $\kappa/\tau = 0$ we have a straight line, when $\kappa/\tau = \infty$, a plane curve.

If we project the helix $\mathbf{x}(s)$ on a plane perpendicular to $\mathbf{a}$, the projection $\mathbf{x}_1$ has the equation (see Fig. 1–30):

$$\mathbf{x}_1 = \mathbf{x} - (\mathbf{x} \cdot \mathbf{a})\mathbf{a}.$$

Hence

$$\mathbf{x}_1' = \mathbf{t} - (\mathbf{t} \cdot \mathbf{a})\mathbf{a} = \mathbf{t} - \mathbf{a} \cos \alpha,$$

and its arc length is given by

$$ds_1^2 = d\mathbf{x}_1 \cdot d\mathbf{x}_1 = \sin^2 \alpha \, ds^2.$$

Since

$$d\mathbf{x}_1/ds_1 = \csc \alpha \mathbf{t} - \cot \alpha \, \mathbf{a},$$

its curvature vector is

$$d^2\mathbf{x}_1/ds_1^2 = \kappa \csc^2 \alpha \mathbf{n},$$

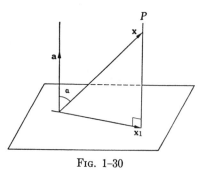

Fig. 1–30

and its curvature $\kappa_1 = \kappa \csc^2 \alpha$. In words:

The projection of a helix on a plane perpendicular to its axis has its principal normal parallel to the corresponding principal normal of the helix, and its corresponding curvature is $\kappa_1 = \kappa \csc^2 \alpha$.

EXAMPLES. (1) *Circular helix.* If a helix has constant curvature, then its projection on a plane perpendicular to its axis is a plane curve of constant curvature, hence a circle (Section 1–8). The helix lies on a cylinder of revolution and is therefore a circular helix.

(2) *Spherical helices.* If a helix lies on a sphere of radius r, then Eq. (8–10) holds, which, together with $\kappa = \tau \tan \alpha$, gives after elimination of τ,

$$r^2 = R^2[1 + R'^2 \tan^2 \alpha],$$

or

$$\frac{R\,dR}{\sqrt{r^2 - R^2}} = \pm ds \cot \alpha,$$

which, integrated for R, and by suitable choice of the additive constant in s, gives

$$R^2 + s^2 \cot^2 \alpha = r^2.$$

The projection of the helix on a plane perpendicular to its axis is therefore a plane curve with the natural equation

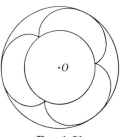

Fig. 1–31

$$R_1^2 + s_1^2 \cos^2 \alpha = r^2 \sin^4 \alpha.$$

This type of curve is discussed in Sec. 1–8, and since $\cos^2 \alpha < 1$, represents an epicycloid (compare Eq. (8–3)):

A spherical helix projects on a plane perpendicular to its axis in an arc of an epicycloid.

This projection is a closed curve when the ratio a/b of Eq. (8–4) is rational. Using the notation of this formula and of Eq. (8–3), we obtain that in this case

$$\cos \alpha = \frac{B}{A} = \frac{a}{a + 2b}$$

is rational.
From

$$B = \frac{4b(a + b)}{a + 2b} = r \sin^2 \alpha,$$

$$A = \frac{4b(a + b)}{a} = r \sin \alpha \tan \alpha,$$

we find

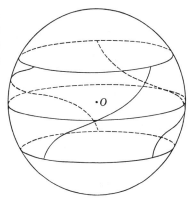

Fig. 1–32.   Only three of the six arcs are drawn.

$$a = r \cos \alpha, \qquad b = \frac{r}{2}(1 - \cos \alpha) = r \sin^2 \alpha/2, \qquad (a + 2b = r)$$

for the radius of the fixed and of the rolling circle. When $\cos \alpha = \frac{3}{5}$ we find as projection an epicycloid of three cusps (Fig. 1–31); each of the three arcs represents part of a helix on the sphere (Fig. 1–32).

This case has been investigated by W. Blaschke, *Monatshefte für Mathematik und Physik* **19**, 1908, pp. 188–204; see also his *Differentialgeometrie*, p. 41.

1–10 General solution of the natural equations. We have seen that the natural equation of a plane curve can be solved (that is, the cartesian coordinates can be found) by means of two quadratures. In the case of a space curve we can try to solve the third order differential equation in $\alpha(s)$ obtained from Eq. (8–6) by eliminating β and γ. The solution can, however, be reduced to that of a first order differential equation, a so-called *Riccati equation*, which is a thoroughly studied type, but not a type which can be solved by quadratures only. This method, due to S. Lie and G. Darboux, is based on the remark that

$$\alpha^2 + \beta^2 + \gamma^2 = 1 \tag{10–1}$$

can be decomposed as follows:

$$(\alpha + i\beta)(\alpha - i\beta) = (1 + \gamma)(1 - \gamma).$$

Let us now introduce the conjugate imaginary functions w and $-z^{-1}$:

$$w = \frac{\alpha + i\beta}{1 - \gamma} = \frac{1 + \gamma}{\alpha - i\beta}; \qquad -\frac{1}{z} = \frac{\alpha - i\beta}{1 - \gamma} = \frac{1 + \gamma}{\alpha + i\beta}. \tag{10–2}$$

Then it is possible to express α, β, γ in terms of w and z:

$$\alpha = \frac{1 - wz}{w - z}, \qquad \beta = i\,\frac{1 + wz}{w - z}, \qquad \gamma = \frac{w + z}{w - z}. \tag{10–3}$$

These expressions are equivalent to the equations of the sphere (Eq. (10–1)) in terms of two parameters w and z. We shall return to this in Sec. 2–3, Exercise 2.

With the aid of the Eqs. (8–6) we find for w':

$$\frac{dw}{ds} = \frac{\alpha' + i\beta'}{1 - \gamma} + \frac{\alpha + i\beta}{(1 - \gamma)^2}\,\gamma' = \frac{\kappa\beta - i\kappa\alpha + i\tau\gamma}{1 - \gamma} + \frac{w\gamma'}{1 - \gamma} = -i\kappa w + \frac{\tau(i\gamma - \beta w)}{1 - \gamma}.$$

Because of Eq. (10–2) we find for β:

$$\beta = i\,\frac{1 + \gamma - w^2 + \gamma w^2}{2w}.$$

The elimination of β from these two equations gives an equation from which γ also has disappeared:

$$\frac{dw}{ds} = -\frac{i}{2}\,\tau - i\kappa w + \frac{i\tau}{2}\,w^2. \tag{10–4}$$

Performing the same type of elimination for dz/ds, we find that z satisfies the same equation as w. This equation has the form

$$\frac{df}{ds} = A + Bf + Cf^2, \text{ where } A(s) = -\frac{i\tau}{2}, B(s) = -i\kappa, C(s) = \frac{i\tau}{2}. \quad (10\text{-}5)$$

This is a so-called *Riccati equation.* It can be shown that its general solution is of the form

$$f = \frac{cf_1 + f_2}{cf_3 + f_4}, \quad (10\text{-}6)$$

where c is an arbitrary constant and the f_i are functions of s.

The fundamental properties of a Riccati equation are:
(1) When one particular integral is known, the general integral can be obtained by two quadratures.
(2) When two particular integrals are known, the general integral can be found by one quadrature.
(3) When three particular integrals f_1, f_2, f_3 are known, every other integral f satisfies the equation

$$\frac{f - f_1}{f - f_2} : \frac{f_3 - f_1}{f_3 - f_2} = c = \text{constant}. \quad (10\text{-}7)$$

In words: The cross ratio of four particular integrals is constant.* Eq. (10-6) is a direct consequence of Eq. (10-7).

Let f_1, f_2, f_3, f_4 be such functions of s that Eq. (10-6) is the general integral of Eq. (10-4). In order to find $\alpha_i, \beta_i, \gamma_i$, $i = 1, 2, 3$, we need three integrals w_i and three integrals z_i, to be characterized by constants c_1, c_2, c_3 for the w_i, and three constants d_1, d_2, d_3 for the z_i:

$$\alpha_1 = \frac{1 - w_1 z_1}{w_1 - z_1}, \qquad \beta_1 = i\frac{1 + w_1 z_1}{w_1 - z_1}, \qquad \gamma_1 = \frac{w_1 + z_1}{w_1 - z_1}, \text{ etc.}$$

The nine functions $\alpha_i, \beta_i, \gamma_i$ must satisfy the orthogonality conditions

$$\alpha_i^2 + \beta_i^2 + \gamma_i^2 = 1, \qquad\qquad i = 1, 2, 3;$$
$$\alpha_i\alpha_j + \beta_i\beta_j + \gamma_i\gamma_j = 0, \qquad i, j = 1, 2, 3, i \neq j.$$

The first three are automatically satisfied by virtue of Eq. (10-1), and therefore we have to find the c_i, d_i in such a way that the last three conditions

---

* Proofs in L. P. Eisenhart, *Differential geometry*, p. 26, or E. L. Ince, *Ordinary differential equations* (1927), p. 23.

are also satisfied. One choice of c_i, d_i is sufficient, since all other choices will give curves congruent to the first choice. Now $\alpha_1\alpha_2 + \beta_1\beta_2 + \gamma_1\gamma_2 = 0$ can be written as follows:

$$(w_1 - w_2)(z_1 - z_2) = -(w_1 - z_2)(z_1 - w_2),$$

or

$$2(w_1z_1 + w_2z_2) = w_1z_2 + w_2z_1 + w_1w_2 + z_1z_2,$$

and if we substitute for w_1, w_2, z_1, z_2 their values (see Eq. (10–6)), with the constants c_1, c_2, d_1, d_2 respectively, we obtain the same relation for the constants:

$$2(c_1d_1 + c_2d_2) = c_1d_2 + c_2d_1 + c_1c_2 + d_1d_2.$$

Similarly:

$$2(c_2d_2 + c_3d_3) = c_2d_3 + c_3d_2 + c_2c_3 + d_2d_3,$$
$$2(c_3d_3 + c_1d_1) = c_3d_1 + c_1d_3 + c_3c_1 + d_3d_1.$$

Every solution of these three equations in six unknowns c_i, d_i will give a coordinate expression for the curve. A simple solution is the following:

$$c_1 = 1, \quad c_2 = i, \quad c_3 = \infty; \qquad d_1 = -1, \quad d_2 = -i, \quad d_3 = 0.$$

To verify this, substitute into the three equations the values of c_1, c_2, d_1, d_2, which gives $c_3d_3 = -1$ as well as $c_3d_3 = +1$ which is compatible with $c_3 = \infty$, $d_3 = 0$. The c and d form three pairs of numbers, each pair of which is harmonic with respect to each other pair. This is a direct result of the properties of the general solution of the Riccati equation. Hence $w_1 = (f_1 + f_2)/(f_3 + f_4)$, etc. We thus obtain for the α_i:

$$\alpha_1 = \frac{1 - w_1z_1}{w_1 - z_1} = \frac{(f_1^2 - f_3^2) - (f_2^2 - f_4^2)}{2(f_1f_4 - f_2f_3)},$$

$$\alpha_2 = \frac{1 - w_2z_2}{w_2 - z_2} = \left[\frac{(f_1^2 - f_3^2) + (f_2^2 - f_4^2)}{2(f_1f_4 - f_2f_3)}\right]i,$$

$$\alpha_3 = \frac{1 - w_3z_3}{w_3 - z_3} = \frac{f_3f_4 - f_1f_2}{f_1f_4 - f_2f_3}, \qquad (10\text{–}8)$$

which results in the following theorem.

If the general solution of the Riccati equation

$$\frac{df}{ds} = -\frac{i}{2}\tau - i\kappa f + \frac{i\tau}{2}f^2$$

is found in the form (f_1, f_2, f_3, f_4 *functions of* s)

$$f = \frac{cf_1 + f_2}{cf_3 + f_4}, \qquad \text{(c arbitrary constant)}$$

then the curve given by the equations

$$x = \int^s \alpha_1 \, ds, \qquad y = \int^s \alpha_2 \, ds, \qquad z = \int^s \alpha_3 \, ds,$$

where the α_i are given by Eq. (10–8), has $\kappa(s)$ and $\tau(s)$ as curvature and torsion.

This reduction to a Riccati equation goes in principle back to Sophus Lie (1882, *Werke III*, p. 531) and was fully established by G. Darboux, *Leçons I*, Ch. 2. We find Eq. (10–8) in G. Scheffers, *Anwendung I*, p. 298.

EXAMPLES. *Plane curve.* When $\tau = 0$:

$$df/f = -i\kappa \, ds,$$
$$f = ce^{-i\varphi}, \qquad \varphi = \int \kappa \, ds,$$
$$f_1 = e^{-i\varphi}, \qquad f_2 = 0, \qquad f_3 = 0, \qquad f_4 = 1,$$

which lead to the Eqs. (8–2) of the plane curve.

Cylindrical helix. In this case the Riccati equation (10–4) can be written in the form

$$w' = -\tfrac{1}{2}\tau i(1 + 2cw - w^2). \qquad (c \text{ constant})$$

Two integrals can immediately be found by taking $w^2 - 2cw + 1 = 0$. The general solution of this equation can now be found by means of one quadrature. For the details of this problem we may refer to Eisenhart, *Differential geometry*, p. 28.

1–11 Evolutes and involutes. The tangents to a space curve $\mathbf{x}(s)$ generate a surface. The curves on this surface which intersect the generating tangent lines at right angles form the *involutes* (German: *Evolvente;* French: *développantes*) of the curve. Their equation is of the form (Fig. 1–33):

$$\mathbf{y} = \mathbf{x} + \lambda\mathbf{t}. \qquad (\lambda \text{ a function of } s)$$
$$(11\text{–}1)$$

The vector $d\mathbf{y}/ds$ is a tangent vector to the involute. Hence:

$$\mathbf{t} \cdot d\mathbf{y}/ds = 0,$$
$$\mathbf{t} \cdot \left(\mathbf{t} + \lambda\kappa\mathbf{n} + \mathbf{t}\frac{d\lambda}{ds}\right) = 0,$$
$$1 + \frac{d\lambda}{ds} = 0,$$
$$\lambda = \text{const} - s = c - s.$$

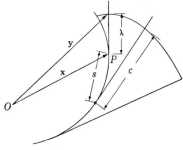

FIG. 1–33

The equation of the involutes is therefore

$$\mathbf{y} = \mathbf{x} + (c - s)\mathbf{t}. \tag{11-2}$$

For each value of c there is an involute; they can be obtained by unwinding a thread originally stretched along the curve, keeping the thread taut all the time.

The converse problem is somewhat more complicated: Find the curves which admit a given curve C as involute. Such curves are called *evolutes* of C (German: *Evolute;* French: *développées*). Their tangents are normal to $C(\mathbf{x})$ and we can therefore write for the equation of the evolute $\mathbf{y}$ (Fig. 1–34):

$$\mathbf{y} = \mathbf{x} + a_1\mathbf{n} + a_2\mathbf{b}.$$

Hence

$$\frac{d\mathbf{y}}{ds} = \mathbf{t}(1 - a_1\kappa) + \mathbf{n}\left(\frac{da_1}{ds} - \tau a_2\right)$$
$$+ \mathbf{b}\left(\frac{da_2}{ds} + \tau a_1\right)$$

must have the direction of $a_1\mathbf{n} + a_2\mathbf{b}$, the tangent to the evolute:

$$\kappa = 1/a_1, \qquad R = a_1,$$

and

$$\frac{\dfrac{da_1}{ds} - \tau a_2}{a_1} = \frac{\dfrac{da_2}{ds} + \tau a_1}{a_2},$$

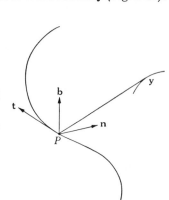

Fig. 1–34

which can be written in the form

$$\frac{a_2 \dfrac{dR}{ds} - R \dfrac{da_2}{ds}}{a_2^2 + R^2} = \tau.$$

This expression can be integrated:

$$\tan^{-1}\frac{R}{a_2} = \int \tau \, ds + \text{const},$$

or

$$a_2 = R\left[\cot\left(\int \tau \, ds + \text{const}\right)\right].$$

The equation of the evolutes is:

$$\mathbf{y} = \mathbf{x} +$$
$$R\left[\mathbf{n} + \cot\left(\int \tau \, ds + \text{const}\right)\mathbf{b}\right]. \tag{11-3}$$

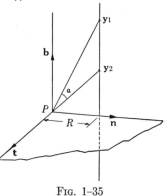

Fig. 1–35

This equation shows that *a point of the evolute lies on the polar axis of the corresponding point of the curve, and the angle under which two different evolutes are seen from the given curve is constant* (Fig. 1–35).

For plane curves:

$$\mathbf{y} = \mathbf{x} + R\mathbf{n} + \lambda R\mathbf{b}. \qquad (\lambda \text{ a constant})$$

For $\lambda = 0$ we obtain the plane evolute. The other evolutes lie on the cylinder erected on this plane evolute as base and with generating lines perpendicular to the plane. They are helices on this cylinder.

The theory of space evolutes is due to Monge (1771, published in 1785, the paper is reprinted in the *Applications*). Further studies were published by Lancret, *Mémoires présentées à l'Institut*, Paris, **1**, 1805, and **2**, 1811. Lancret discussed the "développoïdes" of a curve, which are the curves of which the tangents intersect the curve at constant angle not $= 90°$.

The locus of the polar axes is a surface called the *polar developable* (see Sec. 2–4). On this surface lie the ∞^1 evolutes of the curve and also the locus of the osculating spheres. This locus cannot be one of the evolutes, since its tangent has the direction of the binormal $\mathbf{b}$ (see Eq. (8–11)), while the tangent to the evolute has the direction $R\mathbf{n} + a_2\mathbf{b}$.

EXERCISES

1. The perpendicular distance d of a point $Q(\mathbf{y})$ to a line passing through $P(\mathbf{x})$ in the direction of the unit vector $\mathbf{u}$ is $d = |(\mathbf{y} - \mathbf{x}) \times \mathbf{u}|$ (Fig. 1–36). Using this formula, show that the tangent has a contact of order one with the curve.

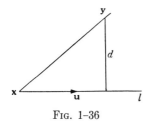

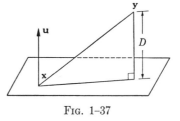

FIG. 1–36   FIG. 1–37

2. The perpendicular distance D of a point $Q(\mathbf{y})$ to a plane passing through $P(\mathbf{x})$ and perpendicular to the unit vector $\mathbf{u}$ is $D = |(\mathbf{y} - \mathbf{x}) \cdot \mathbf{u}|$ (Fig. 1–37). Using this formula, show that a plane through the tangent has a contact of order one, but for the osculating plane which has a contact of order two with the curve.

3. Show (a) that the tangent has a contact of order n with the curve, if $\mathbf{x}'$, $\mathbf{x}''$, $\mathbf{x}'''$, ... $\mathbf{x}^{(n)}$, but not $\mathbf{x}^{(n+1)}$, have the direction of the tangent; and (b) that the osculating plane has such a contact, if $\mathbf{x}'$, $\mathbf{x}''$, $\mathbf{x}'''$, ... $\mathbf{x}^{(n)}$, but not $\mathbf{x}^{(n+1)}$ lie in the osculating plane.

4. Show that the distance of a point $P(\mathbf{x}_1)$ to the plane $\mathbf{a} \cdot \mathbf{x} + p = 0$ is of the order of $\mathbf{a} \cdot \mathbf{x}_1 + p$, and that of P to the sphere $(\mathbf{x} - \mathbf{a}) \cdot (\mathbf{x} - \mathbf{a}) - r^2 = 0$ is of order $(\mathbf{x}_1 - \mathbf{a}) \cdot (\mathbf{x}_1 - \mathbf{a}) - r^2$.

5. *Osculating helix.* This is the circular helix passing through a point P of the curve C, having the same tangent, curvature vector, and torsion. Show that its contact with the curve is of order two. Is it the only circular helix which has a contact of order two with C at P? (T. Olivier, *Journal École Polytechn.*, cah. **24**, tome 15, 1835, pp. 61–91, 252–263.) Also show that the axis of the osculating helix is the limiting position of the common perpendicular of two consecutive principal normals.

6. Starting with the common coordinate equations, find the natural equations of

(a) logarithmic spiral:

$$r = ce^{k\theta}.$$

(b) cycloid:

$$x = a(\theta - \sin\theta), \qquad y = a(1 - \cos\theta).$$

(c) circle involute:

$$x = a(\cos\theta + \theta\sin\theta), \qquad y = a(\sin\theta - \theta\cos\theta).$$

(d) catenary:

$$y = (a/2)(e^{x/a} + e^{-x/a}).$$

7. Find from Eq. (8–6), in the case of constant κ and τ, the third order equation for $\boldsymbol{\alpha}$, and by integration obtain the circular helix.

8. Prove that when the twisted cubic

$$x = at, \qquad y = bt^2, \qquad z = t^3$$

satisfies the equation $2b^2 = 3a$, it is a helix on a cylinder with generating line parallel to the XOZ-plane, making with the X-axis an angle of $45°$. Determine the equation of the cylinder.

9. The spherical indicatrix of a curve is a circle if and only if the curve is a helix.

10. The tangent to the locus C of the centers of the osculating circles of a plane curve has the direction of the principal normal of the curve; its arc length between two of its points is equal to the difference of the radii of curvature of the curve at these points.

11. *Curves of Bertrand.* When a curve C_1 can be brought into a point-to-point correspondence with another curve C so that at corresponding points P_1, P the curves have the same principal normal, then

(a) P_1P is constant $= a$,

(b) the tangent to C at P and the tangent to C_1 at P_1 make a constant angle α,

(c) there exists for C (and similarly for C_1) a linear relation between curvature and torsion

$$\kappa + \tau \cot\alpha = 1/a, \qquad \left(\text{take } \tau \neq 0, \quad \alpha \neq 0, \quad \alpha \neq \frac{\pi}{2}\right) \quad \text{(A)}$$

These curves were first investigated by J. Bertrand (*Journal de Mathém.* **15**, 1850, pp. 332–350).

12. Show that a curve C for which there exists a linear relation between curvature and torsion:

$$P\kappa + Q\tau = R, \qquad (P, Q, R \text{ constants} \neq 0)$$

admits a Bertrand mate, that is, a curve of which the points can be brought into a one-to-one correspondence with the points of C such that at corresponding points the curves have the same principal normal.

13. *Special cases of Bertrand curves.* Show:

(a) When $\tau = 0$ every curve has an infinity of Bertrand mates, $\alpha = 0$.

(b) When $\tau \neq 0$, $\alpha = \pi/2$, we have curves of constant curvature. Each of the curves C and C_1 is the locus of the centers of curvature of the other curve.

(c) When κ and τ are both constant, hence in the case of a circular helix, there are an infinite number of Bertrand mates, all circular helices.

14. Show that the equation of a Bertrand curve (that is, a curve for which (A), Exercise 11, holds) can be written in the form

$$\mathbf{x} = a \int \mathbf{u} \, d\sigma + a \cot \alpha \int \mathbf{u} \times d\mathbf{u}, \qquad (\alpha, a \text{ constants})$$

Here $\mathbf{u} = \mathbf{u}(\sigma)$ is an arbitrary curve on the unit sphere referred to its arc length (hence $\mathbf{u} \cdot \mathbf{u} = 1, \mathbf{u}' \cdot \mathbf{u}' = 1, \mathbf{u}' = d\mathbf{u}/d\sigma$). (Darboux, *Leçons I*, pp. 42–45.) Show that the first term alone on the right-hand side gives curves of constant curvature, the second term alone curves of constant torsion.

15. *Mannheim's theorem.* If P and P_1 are corresponding points on two Bertrand mates, and C and C_1 their centers of curvature, then the cross ratio $(CC_1, PP_1) = \sec^2 \alpha = \text{const.}$

16. Investigate the pairs of curves C_1 and C which can be brought into a point-to-point correspondence such that at corresponding points (a) the tangents are the same, (b) the binormals are the same.

See E. Salkowski, *Math. Annalen* **66**, 1909, pp. 517–557, A. Voss, *Sitzungsber. Akad. München*, 1909, 106 pp., where also other pairs of corresponding curves are discussed.

17. Show that when the curve $\mathbf{x} = \mathbf{x}(s)$ has constant torsion τ, the curve

$$\mathbf{y} = - T\mathbf{n} + \int \mathbf{b} \, ds$$

has constant curvature $\pm \tau$.

18. Show that if in Sec. 1–10 we split $\alpha^2 + \beta^2 + \gamma^2 = 1$ into $(\alpha + i\gamma)(\alpha - i\gamma) = (1 + \beta)(1 - \beta)$ and follow the method indicated by Eq. (10–2), we are led to the Riccati equation

$$w' = -\tfrac{1}{2}(\kappa - i\tau) + \tfrac{1}{2}(\kappa + i\tau)w^2.$$

19. *Loxodromes.* These can be defined as curves which intersect a pencil of planes at constant angle α. Show that their equation can be brought into the form

$$x = r \cos \theta, \qquad y = r \sin \theta, \qquad z = f(r), \qquad \text{where } r = \sqrt{x^2 + y^2}$$

and

$$\theta = \tan \alpha \int \sqrt{1 + (f')^2} \, \frac{dr}{r}.$$

20. The tangents to a helix intersect a plane perpendicular to its projecting cylinder in the points of the involute of the base of the cylinder.

21. Find the involutes of a helix.

22. Show that the helices on a cone of revolution project on a plane perpendicular to their axes (the base) as logarithmic spirals and then show that the intrinsic equations of these conical helices are

$$R = as, \qquad T = bs, \qquad\qquad (a, b \text{ constants})$$

23. Show that the helices on a paraboloid of revolution project on a plane perpendicular to the axis as circle involutes.

24. A necessary and sufficient condition that a curve be a helix is that

$$(\mathbf{x}^{(iv)}\mathbf{x}'''\mathbf{x}'') = -\kappa^5 \frac{d}{ds}\left(\frac{\tau}{\kappa}\right) = 0.$$

25. *Differential equation of space curves.* If, instead of using the method of the Riccati equation, we attempt to obtain the $\mathbf{x}(s)$ directly as the solution of the Frenet equation for given κ and τ, we obtain the equation $(\kappa, \tau \neq 0)$

$$\mathbf{x}^{(iv)} - \left(2\frac{\kappa'}{\kappa} + \frac{\tau'}{\tau}\right)\mathbf{x}''' + \left(\kappa^2 + \tau^2 - \frac{\kappa\kappa'' - 2(\kappa')^2}{\kappa^2} + \frac{\kappa'\tau'}{\kappa\tau}\right)\mathbf{x}'' + \kappa^2\left(\frac{\kappa'}{\kappa} - \frac{\tau'}{\tau}\right)\mathbf{x}' = 0.$$

26. The parabolas $y = x^2$ and $y = \sqrt{x}$ can be obtained from each other by a rotation of 90°. They must therefore satisfy the same natural equation. Obtain this equation in the form

$$\left(\frac{dR}{ds}\right)^2 = 9[\sqrt[3]{4R^2} - 1].$$

27. Verify by using Eq. (8–12) that the curve of Exercise 9, Section 1–6:

$$x = a \sin^2 u, \qquad y = a \sin u \cos u, \qquad z = a \cos u,$$

lies on a sphere.

1–12 Imaginary curves. We have so far admitted only real curves, defined by real functions of a real variable. When we admit complex analytic functions x_i of a complex variable u, then we obtain structures of ∞^2 points, called *imaginary curves.* The formulas for the arc length, tangent, osculating plane, curvature, and torsion retain a formal meaning for those curves, but now they serve as definition of these concepts. Certain theorems require modification, notably those based on the assumption that $ds^2 > 0$. We have to admit, in particular, curves for which $ds^2 = 0$, the *isotropic curves.* And if we define *planes* as structures for which a linear relation exists between the x_i, or — what amounts to the same — as *plane curves,* curves for which $(\dot{\mathbf{x}}\,\ddot{\mathbf{x}}\,\dddot{\mathbf{x}}) = 0$, then we have to admit planes such as the plane with equation $x_2 = ix_1$, for which ds^2 can no longer be written as the sum of the squares of two differentials, but as the square of a single differential such as $ds^2 = dx_3^2$ in the case of $x_2 = ix_1$. We therefore distinguish between *regular planes*, of which the equation by proper choice of the coordinates can be reduced to $x_3 = 0$, and *isotropic planes*, of which the equation can be reduced to $x_2 = ix_1$. By *planes* we mean regular planes. There are no planes for which $ds^2 = 0$. (Such planes do exist in four-space; e.g., those for which $x_1 = ix_2$, $x_3 = ix_4$.)

Isotropic curves in the regular plane $x_3 = 0$ are defined by

$$ds^2 = dx_1^2 + dx_2^2 = 0; \qquad dx_2 = \pm i\, dx_1, \tag{12-1}$$

which equation gives, by integration:

$$x_2 - b = \pm i(x_1 - a). \tag{12-2}$$

In an isotropic plane we obtain by integration of $ds^2 = dx_3^2 = 0$ that $x_3 = c$, $x_2 = ix_1$. Hence:

Plane isotropic curves are straight lines. Through every point of a regular plane pass two isotropic lines; through every point of an isotropic plane passes one isotropic line. The isotropic lines in a regular plane form two sets of parallel lines, those in an isotropic plane form one set of parallel lines.

Isotropic lines were introduced by V. Poncelet in his *Traité des propriétés projectives des figures* (1822). They are also called *minimal curves* or *null curves*.

Isotropic curves in space $\mathbf{x} = \mathbf{x}(u)$ satisfy the differential equation

$$d\mathbf{x} \cdot d\mathbf{x} = dx_1^2 + dx_2^2 + dx_3^2 = 0, \tag{12-3}$$

or

$$\dot{\mathbf{x}} \cdot \dot{\mathbf{x}} = 0, \qquad \dot{\mathbf{x}} = d\mathbf{x}/du. \tag{12-4}$$

One solution of this differential equation consists of the isotropic straight lines

$$\mathbf{X} = \mathbf{x} + u\mathbf{a}, \qquad \mathbf{x, a} \text{ constant vectors, } \mathbf{a} \cdot \mathbf{a} = 0, \tag{12-5}$$

for which, as we see by differentiation, $d\mathbf{X} \cdot d\mathbf{X} = \mathbf{a} \cdot \mathbf{a}\, du^2 = 0$. These lines generate, at each point $P(\mathbf{x})$, a quadratic cone

$$(\mathbf{X} - \mathbf{x}) \cdot (\mathbf{X} - \mathbf{x}) = 0, \tag{12-6}$$

the *isotropic cone* with vertex P. The tangent plane to this cone along the generator (12-5) has the equation

$$(\mathbf{X} - \mathbf{x}) \cdot \mathbf{a} = 0, \qquad \mathbf{a} \cdot \mathbf{a} = 0. \tag{12-7}$$

This is easily verified if we remember that the tangent plane to the cone $x^2 + y^2 + z^2 = 0$ at (x_1, y_1, z_1) is $xx_1 + yy_1 + zz_1 = 0$, and that this plane remains the same when we multiply x_1, y_1, z_1 by a factor λ, so that this plane is tangent along the generator $x : y : z = x_1 : y_1 : z_1 = a_1 : a_2 : a_3$.

By selecting $\mathbf{x} = 0$ and taking the XOY-plane through $\mathbf{a}$, the equation of plane (12–7) becomes $x_1 = \pm ix_2$, so that (12–7) is an isotropic plane. Eq. (12–7) is, at the same time, the most general form of the equation of an isotropic plane, since $\mathbf{a} \cdot \mathbf{a} = 0$ for the planes $x_1 = \pm ix_2$ and therefore also zero for all positions of a rectangular cartesian coordinate system.

Substitution of the $\mathbf{X}$ of Eq. (12–5) into Eq. (12–7) shows that isotropic line (12–5) lies in the plane (12–7).   Hence:

The tangent planes to the isotropic cones are isotropic planes, tangent along isotropic lines, and also, although it seems strange at first:

The normal to an isotropic plane at a point in this plane lies in this plane and is tangent to the isotropic cone at that point.

Isotropic straight lines do not form the only solution of Eq. (12–4). We shall now derive the general solution by a method reminiscent of that used to pass from Eq. (10–1) to Eq. (10–2).   For this purpose we write Eq. (12–3) in the form

$$\frac{dx_1 + i\, dx_2}{dx_3} = -\frac{dx_3}{dx_1 - i\, dx_2} = u.$$

This leads to the equations

$$\frac{dx_1}{dx_3} + i\frac{dx_2}{dx_3} = u, \qquad \frac{dx_1}{dx_3} - i\frac{dx_2}{dx_3} = -\frac{1}{u}.$$

Solving these two equations for dx_1/dx_3 and dx_2/dx_3, we obtain the solution of Eq. (12–4) in the form

$$\frac{dx_1}{u^2 - 1} = \frac{dx_2}{i(u^2 + 1)} = \frac{dx_3}{2u} = F(u)\, du.$$

Writing $f'''(u)$ for $F(u)$, *we then obtain by partial integration the equation of the isotropic curves in the form:*

$$\begin{aligned}
x_1 &= (u^2 - 1)f'' - 2uf' + 2f, \\
x_2 &= [(u^2 + 1)f'' - 2uf' + 2f]i, \\
x_3 &= 2uf'' - 2f',
\end{aligned} \tag{12–8}$$

where $f(u)$ is an arbitrary analytic function of a parameter u.   This representation contains, as we see, no integration.   It does not give the (straight) isotropic lines.

The tangent line to an isotropic curve at $P(\mathbf{x})$ is given by

$$\mathbf{y} = \mathbf{x} + \lambda\dot{\mathbf{x}}, \tag{12–9}$$

the osculating plane by

$$\mathbf{y} = \mathbf{x} + \lambda\dot{\mathbf{x}} + \mu\ddot{\mathbf{x}}, \quad \text{or} \quad (\mathbf{y} - \mathbf{x}, \dot{\mathbf{x}}, \ddot{\mathbf{x}}) = 0. \tag{12–10}$$

From Eq. (12–4) follows

$$\dot{\mathbf{x}} \cdot \ddot{\mathbf{x}} = 0. \tag{12–11}$$

Since

$$(\dot{\mathbf{x}} \times \ddot{\mathbf{x}}) \cdot (\dot{\mathbf{x}} \times \ddot{\mathbf{x}}) = (\dot{\mathbf{x}} \cdot \dot{\mathbf{x}})(\ddot{\mathbf{x}} \cdot \ddot{\mathbf{x}}) - (\dot{\mathbf{x}} \cdot \ddot{\mathbf{x}})^2, \tag{12–12}$$

we see that

$$(\dot{\mathbf{x}} \times \ddot{\mathbf{x}}) \cdot (\dot{\mathbf{x}} \times \ddot{\mathbf{x}}) = 0. \tag{12–13}$$

Hence, comparing with Eq. (12–7), we find that *the osculating plane of an isotropic curve (not a straight line) is tangent to the isotropic cone.*

However, all curves whose osculating planes are tangent to the isotropic cone are not all isotropic curves. Such curves are characterized by Eq. (12–13). In this case we apply the identity

$$(\mathbf{a} \cdot \mathbf{a})(\mathbf{abc})^2 = [(\mathbf{a} \times \mathbf{b}) \cdot (\mathbf{a} \times \mathbf{b})][(\mathbf{a} \times \mathbf{c}) \cdot (\mathbf{a} \times \mathbf{c})] - [(\mathbf{a} \times \mathbf{b}) \cdot (\mathbf{a} \times \mathbf{c})]^2. \tag{12–14}$$

Identity (12–14) is obtained by substituting $\mathbf{p} = \mathbf{a} \times \mathbf{b}$, $\mathbf{q} = \mathbf{a} \times \mathbf{c}$ into the identity

$$(\mathbf{p} \times \mathbf{q}) \cdot (\mathbf{p} \times \mathbf{q}) = (\mathbf{p} \cdot \mathbf{p})(\mathbf{q} \cdot \mathbf{q}) - (\mathbf{p} \cdot \mathbf{q})^2,$$

and applying to $\mathbf{p} \times \mathbf{q}$ the identity

$$(\mathbf{u} \times \mathbf{v}) \times \mathbf{w} = (\mathbf{u} \cdot \mathbf{w})\mathbf{v} - (\mathbf{v} \cdot \mathbf{w})\mathbf{u},$$

or

$$(\mathbf{a} \times \mathbf{b}) \times (\mathbf{a} \times \mathbf{c}) = -[\mathbf{b} \cdot (\mathbf{a} \times \mathbf{c})]\mathbf{a} = (\mathbf{abc})\mathbf{a}.$$

When in this identity $\mathbf{a} = \dot{\mathbf{x}}$, $\mathbf{b} = \ddot{\mathbf{x}}$, $\mathbf{c} = \dddot{\mathbf{x}}$, then we obtain for the case that Eq. (12–13) is satisfied,

$$(\dot{\mathbf{x}} \cdot \dot{\mathbf{x}})(\dot{\mathbf{x}} \ddot{\mathbf{x}} \dddot{\mathbf{x}})^2 = [(\dot{\mathbf{x}} \times \ddot{\mathbf{x}}) \cdot (\dot{\mathbf{x}} \times \dddot{\mathbf{x}})]^2 = 0, \tag{12–15}$$

since differentiation of Eq. (12–13) gives

$$(\dot{\mathbf{x}} \times \ddot{\mathbf{x}}) \cdot (\dot{\mathbf{x}} \times \dddot{\mathbf{x}}) = 0.$$

Eq. (12–15) expresses the *theorem of E. Study:*

When at every point of a curve the osculating plane is an isotropic plane the curve is either an isotropic curve or a curve in an isotropic plane.

1–13 Ovals. We shall now present some material which no longer belongs to *local* differential geometry, but to *differential geometry in the large.* Theorems in this field do not deal with the exclusive behavior of curves and surfaces in the immediate neighborhood of a point, such as the curvature or torsion "at" a point, but describe characteristics of a finite arc or segment of the curve or surface. Such theorems are often obtained by means of the integral calculus, where local differential geometry is based on the application of the differential calculus. For this reason we sometimes oppose *integral geometry* to *differential geometry*. Classical integration of arc lengths, areas, and volumes can be considered as part of integral geometry. Although there is no reason to attempt a specific distinction between integral geometry and differential geometry in the large, we are inclined to use the latter term when dealing with such conceptions as curvature and torsion, defined in "local" differential geometry.

A section of differential geometry in the large which has been rather well investigated deals with ovals. *An oval is a real, plane, closed, twice differentiable curve of which the curvature vector always points to the interior.* Then the tangent turns continuously to the left when we proceed counterclockwise along the curve. The points can be paired into sets of *opposite* points, at which the tangents are parallel (Fig. 1–38). If $\mathbf{t}$ is the unit tangent vector at P, then $\mathbf{t}_1 = -\mathbf{t}$ is the unit tangent vector at the opposite point P_1. Then the curvature $\kappa = d\varphi/ds$ cannot change its sign, since φ is a monotonically increasing (decreasing) function of s.

It will be remembered that in the case of plane curves it is possible to determine the sign of κ by postulating that the sense of rotation $\mathbf{t} \to \mathbf{n}$ is that of $OX \to OY$. When, therefore, κ maintains the same sign along the curve, the spherical image of the curve is a circle traversed once in the same sense.

We shall assume the curvature $\kappa(s)$ to be differentiable and > 0. A point where κ has an extreme value is called a *vertex*. We now prove the following theorem.

Four-vertex theorem. An oval has at least four vertices.

To prove this theorem we observe that when the oval is not a circle it has at least two vertices A and B. Suppose that there are no more, that

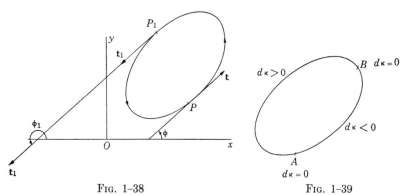

FIG. 1–38   FIG. 1–39

$d\kappa > 0$ on one side of AB and $d\kappa < 0$ on the other side (Fig. 1–39).   Now take $\int \mathbf{n}\, dR$ along the curve ($R = \kappa^{-1}$):

$$\int_C \mathbf{n}\, dR = R\mathbf{n}\Big|_C - \int_C R\, d\mathbf{n} = \int R\kappa \mathbf{t}\, ds = \int \mathbf{t}\, ds = \int d\mathbf{x} = 0.$$

This means that $\int_A^B \mathbf{n}\, dR$ has the same value when taken along the arc AB where $dR > 0$ and along the arc AB where $dR < 0$. If we take the spherical image of $\mathbf{n}$ (Fig. 1–40), in which A_1 and B_1 correspond to A and B, then we see that the two integrals give vectors lying in segments of the plane in which they cannot add up to zero. Therefore there are more than two vertices, and since the curve is closed, their number is even. Four is possible, since an ellipse has four vertices.

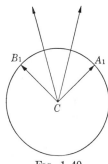

Fig. 1–40

This theorem was found by S. Mukhopadhyaya, *Bull. Calcutta Math. Soc. I* (1909), *Coll. geom. papers I*, pp. 13–20; here a *cyclic point* is defined as a point where the circle of curvature passes through four consecutive points, a *sextactic* point where the osculating conic passes through six consecutive points, etc. The theorem was rediscovered by A. Kneser, *H. Weber Festschrift* 1912, pp. 170–180; several proofs have since been given, e.g., one in W. Blaschke, *Differentialgeometrie*, p. 31; see also S. B. Jackson, *Bull. Amer. Math. Soc.* **50**, 1944, pp. 564–578; P. Scherk, *Proc. First Canadian Math. Congress*, 1945, pp. 97–102.

A convenient set of formulas for ovals can be obtained by associating two points $P(\mathbf{x})$ and $P_1(\mathbf{x}_1)$ of the oval with parallel tangents or opposite points by means of the equations (Fig. 1–41):

$$\mathbf{x}_1 = \mathbf{x} + \lambda\mathbf{t} + \mu\mathbf{n}, \qquad \mathbf{x} = \mathbf{x}(s), \quad (13\text{–}1)$$

where μ is the *width* of the curve at p; that is, the distance between the two parallel tangents at P.   Then, denoting the quantities at P_1 by index 1, we have

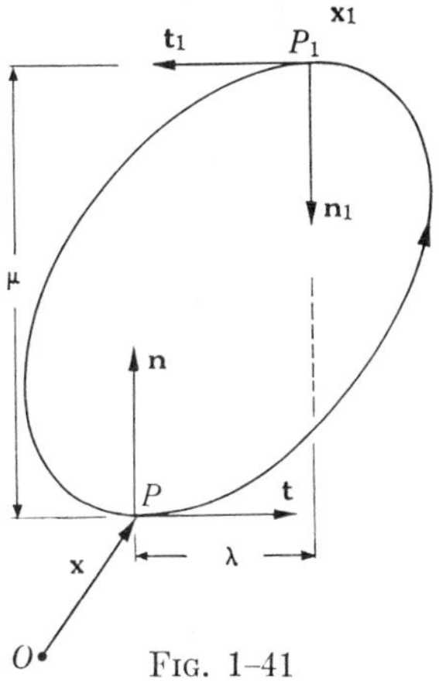

$$\mathbf{t}_1 = -\mathbf{t}, \qquad \mathbf{n}_1 = -\mathbf{n}.$$

Fig. 1–41

Differentiation of Eq. (13–1) gives

$$t_1 \, ds_1/ds = t + \lambda't + \lambda\kappa n + \mu'n - \mu\kappa t,$$

or, equating the coefficients of **t** and **n** on both sides of this equation:

$$1 + ds_1/ds = -\lambda' + \mu\kappa,$$
$$0 = \lambda\kappa + \mu', \qquad (13\text{--}2a)$$

or, in terms of differentials, using the equation $\kappa \, ds = d\varphi$:

$$ds + ds_1 = -d\lambda + \mu \, d\varphi, \qquad 0 = \lambda \, d\varphi + d\mu. \qquad (13\text{--}2b)$$

These equations lead to some simple results in the case of *curves of constant width*, or *orbiform curves*, characterized by the property that the distance between two parallel tangents is always constant. Then $d\mu = 0$. Then Eqs. (13–2) give $\lambda = 0$, and integration from $\varphi = 0$ to $\varphi = \pi$ shows that the perimeter of the curve P is equal to $\mu\pi$:

$$P = \int_{\varphi=0}^{\varphi=\pi} (ds + ds_1) = \mu \int_0^\pi d\varphi = \mu\pi.$$

In words:

The chord connecting opposite points of a curve of constant width is perpendicular to the tangents at these points. And

Theorem of Barbier. All curves of constant width μ have the same perimeter $\mu\pi$.

The first of these theorems establishes not only a necessary but also a sufficient condition.

A circle is a curve of constant width, but there are some curves of constant width which are not circles. This can be shown by following a method indicated by Euler. If we wish to construct a curve of constant width with $2n$ vertices, we take a closed differentiable curve with n cusps having one tangent in every direction (such curves may be called *curves of zero width*). *Any involute of such a curve is a curve of constant width.* Let us, for instance take $n = 3$ and let us start with a point

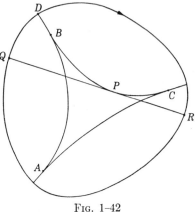

FIG. 1–42

R on the tangent at point P (Fig. 1–42). Then, developing the involute from P via B, A, C back to P, we obtain a closed curve with two tangents

in every direction; the distance between two opposite points Q, R is
$QR = PQ + PR =$ arc $PB + DB + DB +$ arc $AB -$ arc $AC +$ arc $CP =$
arc $BC +$ arc $AB -$ arc $AC + 2DB$. The curvature vector of the involute,
which at Q is along QP, always points to the interior.

Curves of constant width were introduced by L. Euler, *De curvis triangularibus, Acta Acad. Petropol.* 1778 (1780) II, pp. 3–30, who called them *orbiform curves*, and the curve ABC of Fig. 1–42 a *triangular curve*. E. Barbier, *Journal de mathem.* (2) **5** (1860), pp. 272–286, connected the theory of these curves with that of the needle problem in probability. (See further C. Jordan and R. Fiedler, *Contribution à l'étude des courbes convexes fermées* (1912).) The Eqs. (13–2) were first established by A. P. Mellish, *Annals of Mathem.* **32**, 1931, pp. 181–190. For more information on differential geometry in the large see W. Blaschke, *Differentialgeometrie* and *Einführung*; also D. J. Struik, *Bulletin Amer. Mathem. Soc.* **37**, 1931, pp. 49–62.

EXERCISES

1. The limit of the ratio of arc to chord at a point P of a curve is, for chord $\to 0$, unity only when the tangent at P is not isotropic. Prove for the case of a plane curve that this limit is

$$\frac{2\sqrt{k}}{k+1},$$

when $k - 1$ is the order of contact of the curve with the isotropic tangent. (E. Kasner, *Bull. Am. Math. Soc.* **20**, 1913–1914, pp. 524–531; *Proc. Nat. Ac. Sc.* **18**, 1932, pp. 267–274.)

2. Show that the angle of an isotropic straight line with itself is indeterminate.

3. When $\mathbf{a} \cdot \mathbf{a} = 0$, and $\mathbf{a} \cdot \mathbf{b} = 0$, (a) show that $\mathbf{b}$ lies in the isotropic plane through $\mathbf{a}$, (b) that every vector $\mathbf{c}$ for which $\mathbf{c} \cdot \mathbf{a} = 0$ lies in this plane.

4. Isotropic curves are helices on all cylinders passing through them.

5. The equation

$$(X_1 - x_1)\left(\frac{1 - u^2}{2}\right) - i(X_2 - x_2)\left(\frac{1 + u^2}{2}\right) + (X_3 - x_3)u = 0,$$

represents an isotropic plane.

6. By means of the expressions $\dot{\mathbf{x}} \cdot \dot{\mathbf{x}}$, $(\dot{\mathbf{x}}\,\ddot{\mathbf{x}}\,\dddot{\mathbf{x}})$, $\dot{\mathbf{x}} \times \ddot{\mathbf{x}}$, $(\dot{\mathbf{x}} \times \ddot{\mathbf{x}}) \cdot (\dot{\mathbf{x}} \times \ddot{\mathbf{x}})$ characterize: (a) regular straight lines, (b) regular plane curves, (c) isotropic straight lines, (d) isotropic curves, (e) curves in isotropic planes.

(E. Study, *Transactions Am. Mathem. Soc.* **10**, 1909, p. 1.)

What is κ and τ in each case?

(*Example:* For a regular straight line $\dot{\mathbf{x}} \cdot \dot{\mathbf{x}} \neq 0$, $\dot{\mathbf{x}} \times \ddot{\mathbf{x}} = 0$, $(\dot{\mathbf{x}} \times \ddot{\mathbf{x}}) \cdot (\dot{\mathbf{x}} \times \ddot{\mathbf{x}}) = 0$; $(\dot{\mathbf{x}}\,\ddot{\mathbf{x}}\,\dddot{\mathbf{x}}) = 0$.)

7. By using the identity

$$(\dot{\mathbf{x}}\,\ddot{\mathbf{x}}\,\dddot{\mathbf{x}})^2 = \begin{vmatrix} \dot{\mathbf{x}} \cdot \dot{\mathbf{x}} & \dot{\mathbf{x}} \cdot \ddot{\mathbf{x}} & \dot{\mathbf{x}} \cdot \dddot{\mathbf{x}} \\ \ddot{\mathbf{x}} \cdot \dot{\mathbf{x}} & \ddot{\mathbf{x}} \cdot \ddot{\mathbf{x}} & \ddot{\mathbf{x}} \cdot \dddot{\mathbf{x}} \\ \dddot{\mathbf{x}} \cdot \dot{\mathbf{x}} & \dddot{\mathbf{x}} \cdot \ddot{\mathbf{x}} & \dddot{\mathbf{x}} \cdot \dddot{\mathbf{x}} \end{vmatrix},$$

show that a plane isotropic curve is a straight line (E. Study, *loc. cit.*).

8. The sum of the radii of curvature at opposite points of a curve of constant width is equal to the width.

9. The statements:
(a) an oval is of constant width,
(b) an oval has the property that PP_1 is constant (constant *diameter*),
(c) all normals of an oval are double (i.e., are normals at two points),
(d) for an oval the sum of the radii of curvature at opposite points is constant,
are equivalent, in the sense that, whenever one of the statements (a), (b), (c), (d) holds true, all other statements are true (A. P. Mellish).

10. The perimeter of an oval is equal to π times its mean width, the mean taken with respect to the angle φ.

11. Show that for a general oval the width μ satisfies an equation of the form

$$\frac{d^2\mu}{d\varphi^2} + \mu = f(\varphi),$$

where $f(\varphi) = R + R_1$.

12. The curves with equation

$$x^4 + y^4 = 1,$$

have eight vertices, situated on the lines $x = 0$, $y = 0$, $x \pm y = 0$.

13. *Curvature centroid.* This is the center of gravity of a curve if its arc is loaded with a mass density proportional to its curvature. Show that the curvature centroid of an oval and of its evolute are identical.

14. *Pedal curve.* The locus of the points of intersection of the tangents to a curve C and the perpendiculars through a point A on these tangents is the *pedal curve* of C with respect to A. Find the pedal curve of a circle with respect to a point on its circumference.

15. If P is a point on a curve C and Q is the corresponding point of the pedal curve of C with respect to a point A (see Exercise 14), then AQ makes the same angles with the pedal curve which AP makes with C. Hint: If C is given by $\mathbf{x} = \mathbf{x}(s)$, write the pedal curve as $\mathbf{y} = (\mathbf{x} \cdot \mathbf{n})\mathbf{n}$.

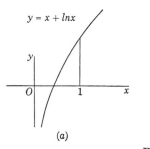

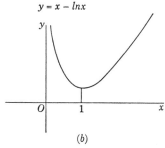

Fig. 1–43

16. The curves $y = x - \ln x$, $y = x + \ln x$ are not congruent (Fig. 1–43). Show that nevertheless both equations have the same natural equation. The reason is that the two curves can pass into each other by an imaginary motion (G. Scheffers, *Anwendung I*, p. 17).

1–14 Monge. We have already had an opportunity to mention some of the contributions of this mathematician, who, with Gauss, can be considered the founder of differential geometry of curves and surfaces. *Gaspard Monge* (1746–1818) started his career as a professor at the military academy in Mézières (on the Meuse, in N. France), where he developed our present descriptive geometry. During the Revolution he was an active Jacobin and occupied leading political and scientific positions; as temporary head of the government on the day of the King's execution he incurred lasting royalist resentment as the chief regicide. After 1795 he became the principal organizer of the Polytechnical School in Paris, the prototype of all our technical institutes and even of West Point. Many leading mathematicians and physicists such as Lagrange and Ampère were connected with this School. Monge was a great teacher, and his lessons in algebraic and differential geometry inspired many younger men; among them were *V. Poncelet*, who established projective geometry, and *C. Dupin*, who contributed greatly to the geometry of surfaces. Other pupils of Monge were *J. B. Meusnier, E. L. Malus, M. A. Lancret*, and *O. Rodrigues*, who all have theorems in differential geometry named after them. Monge's most important papers on the geometry of curves and surfaces have been collected in his *Applications de l'Analyse à la Géometrie* (1807), of which the fifth edition appeared in 1850, with notes of J. Liouville. Monge enjoyed the confidence of Napoleon, and was discharged as director of the Polytechnical School after the fall of the emperor. He died soon afterwards. The main ideas of Chapter 2 are due to Monge and his pupils. Those of Chapter 1 were in part also developed by these men, and in part by a later school of French mathematicians, who associated themselves with the *Journal de Mathématiques pures et appliquées,* continuously published since 1836, when it was founded by *Joseph Liouville* (1809–1882). These authors included *A. J. C. Barré de Saint Venant* (1796–1886), who became interested in the theory of curves through his work in elasticity, and a group of younger men, notably *F. Frenet* (1816–1888), *J. A. Serret* (1819–1885), *V. Puiseux* (1820–1883) and *J. Bertrand* (1822–1900), whose principal papers in this field were written in the period 1840–1850; J. Liouville gave a comprehensive report on these investigations in Note I to the 1850 edition of Monge's *Applications.*

CHAPTER 2

ELEMENTARY THEORY OF SURFACES

2–1 Analytical representation. We shall give a surface, in most cases, by expressing its rectangular coordinates x_i as functions of two parameters u, v in a certain closed interval:

$$x_i = x_i(u, v), \qquad u_1 \leqslant u \leqslant u_2, \qquad v_1 \leqslant v \leqslant v_2. \tag{1–1}$$

The conditions imposed on these functions are analogous to those imposed on the conditions for curves in Section 1–1. We consider the functions x_i to be real functions of the real variables u, v, unless imaginaries are explicitly introduced. When the functions are differentiable to the order $n - 1$, and the nth derivatives exist, we can establish the Taylor formula:

$$x_i(u, v) = x_i(u_0, v_0) + h\left(\frac{\partial x_i}{\partial u}\right)_0 + k\left(\frac{\partial x_i}{\partial v}\right)_0 + \frac{1}{2!}\left(h\frac{\partial}{\partial u} + k\frac{\partial}{\partial v}\right)_0^2 x_i + \cdots$$
$$+ \frac{1}{(n-1)!}\left(h\frac{\partial}{\partial u} + k\frac{\partial}{\partial v}\right)_0^{n-1} x_i + \frac{1}{n!}\left(h\frac{\partial}{\partial u} + k\frac{\partial}{\partial v}\right)_0^{n} x_i(u_0 + \theta h, v_0 + \theta k),$$
$$0 < \theta < 1. \tag{1–2}$$

The parameters u and v must enter independently, which means that the matrix

$$M \equiv \left\|\begin{matrix} \dfrac{\partial x}{\partial u} & \dfrac{\partial y}{\partial u} & \dfrac{\partial z}{\partial u} \\ \dfrac{\partial x}{\partial v} & \dfrac{\partial y}{\partial v} & \dfrac{\partial z}{\partial v} \end{matrix}\right\|, \quad \text{or} \quad \left\|\begin{matrix} x_u & y_u & z_u \\ x_v & y_v & z_v \end{matrix}\right\| \tag{1–3}$$

has rank 2. Points where this rank is 1 or 0 are *singular points;* when the rank at all points is 1 the Eqs. (1–1) represent a curve, as in the case

$$x = u + v, \qquad y = (u + v)^2, \qquad z = (u + v)^3. \tag{1–4}$$

When two determinants of the matrix (1–3) vanish, all three vanish (unless one column contains two zeros), but the vanishing of only one determinant does not mean that the point is singular. If, for example, the surface is given by

$$x = u + v, \qquad y = u + v, \qquad z = uv, \tag{1–5}$$

then $x_u y_v - x_v y_u = 0$, but $x_u z_v - x_v z_u \neq 0$, and the surface is a plane through the Z-axis. Another example is

$$x = f_1(u), \qquad y = f_2(u), \qquad z = v, \tag{1–6}$$

which represents a cylinder.

Singular points may appear because of the nature of the surface and also because of the particular choice of the coordinates. An example of the second case is the *sphere* referred to latitude θ and longitude φ:

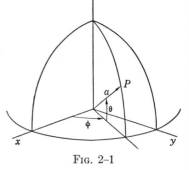

$$x = a \cos \theta \cos \varphi,$$

$$y = a \cos \theta \sin \varphi, \qquad (1\text{-}7)$$

$$z = a \sin \theta,$$

Fig. 2–1

$$M = \left\| \begin{matrix} -a \sin \theta \cos \varphi & -a \sin \theta \sin \varphi & a \cos \theta \\ -a \cos \theta \sin \varphi & +a \cos \theta \cos \varphi & 0 \end{matrix} \right\| \ (a = \text{const}),$$

which has a singular point at $\theta = \dfrac{\pi}{2}$ (the North Pole if the XOY-plane is the equator, Fig. 2–1). This is clearly due to the choice of parameters.

For a *circular cone*, with coordinate representation (Fig. 2–2):

$$x = u \sin \alpha \cos \varphi, \qquad y = u \sin \alpha \sin \varphi, \qquad z = u \cos \alpha, \qquad (1\text{-}8)$$

$$M = \left\| \begin{matrix} \sin \alpha \cos \varphi & \sin \alpha \sin \varphi & \cos \alpha \\ -u \sin \alpha \sin \varphi & u \sin \alpha \cos \varphi & 0 \end{matrix} \right\| \ (\alpha = \text{const}),$$

we find a singular point at $u = 0$, which is the vertex, a particular point of the surface.

When we write the equation of the surface in vector form:

$$\mathbf{x} = \mathbf{x}(u, v) = x_1 \mathbf{e}_1 + x_2 \mathbf{e}_2 + x_3 \mathbf{e}_3, \quad (1\text{-}9)$$

the condition that the rank of matrix M be 2 can be written in the form

$$\mathbf{x}_u \times \mathbf{x}_v \neq 0 \qquad (1\text{-}10)$$
$$(\mathbf{x}_u = \partial \mathbf{x}/\partial u, \ \mathbf{x}_v = \partial \mathbf{x}/\partial v).$$

This equation allows a simple geometrical interpretation. When we keep v constant in Eq. (1–1) or Eq. (1–9), the $\mathbf{x}$ depends on only one parameter u

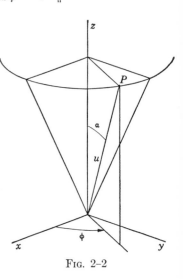

Fig. 2–2

and thus determines a curve on the surface, a *parametric curve*, $v = $ constant. Similarly, $u = $ constant represents another parametric curve. When the constants vary, the surface is covered with a *net* of parametric curves, two of which pass through every point P, forming the *family* of ∞^1
curves $v = $ constant and the *family*
of ∞^1 curves $u = $ constant. At P
the vector $\mathbf{x}_u$ is tangent to the curve
$v = $ constant, and $\mathbf{x}_v$ is tangent to
the curve $u = $ constant (see Section
1–2). Condition (1–10) means that
at P the vectors $\mathbf{x}_u$ and $\mathbf{x}_v$ do not
vanish and have different directions.
(See Fig. 2–3.)

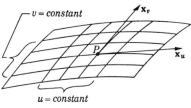

Fig. 2–3

We also call (u, v) the *curvilinear coordinates* of a point on the surface. As curvilinear coordinates of a point on the sphere we may select latitude and longitude, a familiar procedure in geography; polar coordinates are an example of curvilinear coordinates in the plane (rectilinear coordinates can be considered as a special case of curvilinear coordinates). The parametric curves are also called *coordinate curves*.

When we pass from one system of curvilinear coordinates to another

$$u = u(\bar{u}, \bar{v}), \qquad v = v(\bar{u}, \bar{v}), \tag{1–11}$$

we obtain the equation of the surface in the new form

$$\mathbf{x} = \mathbf{x}(\bar{u}, \bar{v}).$$

The tangent vectors to the new parametric lines $\mathbf{x}_{\bar{u}}$ and $\mathbf{x}_{\bar{v}}$ are expressed as follows in terms of $\mathbf{x}_u, \mathbf{x}_v$:

$$\begin{aligned}
\mathbf{x}_{\bar{u}} &= \mathbf{x}_u \frac{\partial u}{\partial \bar{u}} + \mathbf{x}_v \frac{\partial v}{\partial \bar{u}}, \\
\mathbf{x}_{\bar{v}} &= \mathbf{x}_u \frac{\partial u}{\partial \bar{v}} + \mathbf{x}_v \frac{\partial v}{\partial \bar{v}},
\end{aligned} \tag{1–12}$$

so that

$$\mathbf{x}_{\bar{u}} \times \mathbf{x}_{\bar{v}} = \begin{pmatrix} u & v \\ \bar{u} & \bar{v} \end{pmatrix} (\mathbf{x}_u \times \mathbf{x}_v). \tag{1–13}$$

We see that the inequality (1–10) continues to hold for the new system of coordinates, provided the functional determinant

$$\begin{pmatrix} u & v \\ \bar{u} & \bar{v} \end{pmatrix} = \frac{\partial(u, v)}{\partial(\bar{u}, \bar{v})} = \begin{vmatrix} \dfrac{\partial u}{\partial \bar{u}} & \dfrac{\partial v}{\partial \bar{u}} \\ \dfrac{\partial u}{\partial \bar{v}} & \dfrac{\partial v}{\partial \bar{v}} \end{vmatrix} \neq 0. \tag{1–14}$$

We shall consider only coordinate transformations of this kind, which change systems in which $\mathbf{x}_u$ and $\mathbf{x}_v$ have different directions into systems in which $\mathbf{x}_{\bar{u}}$ and $\mathbf{x}_{\bar{v}}$ have different directions, or at any rate restrict our study of the surface to regions where (1–14) holds. Eq. (1–14) can also be interpreted by saying that we consider only those transformations by which independent parameters pass into independent ones in accordance with formula (1–10).

2–2 First fundamental form. A relation $\varphi(u, v) = 0$ between the curvilinear coordinates determines a *curve* on the surface. Such a curve can also be given in parametric form:

$$u = u(t), \qquad v = v(t). \tag{2–1}$$

The vector $dx/dt = \dot{\mathbf{x}}$, at a point P of the surface, given by

$$\dot{\mathbf{x}} = \mathbf{x}_u \dot{u} + \mathbf{x}_v \dot{v}, \tag{2–2}$$

is tangent to the curve and therefore to the surface. Eq. (2–2) can be written in a form independent of the choice of parameter:

$$d\mathbf{x} = \mathbf{x}_u \, du + \mathbf{x}_v \, dv. \tag{2–3}$$

When the curve is given by $\varphi(u, v) = 0$, the du and dv are connected by the relation

$$\varphi_u \, du + \varphi_v \, dv = 0. \tag{2–4}$$

The ratio $dv/du = -\varphi_u/\varphi_v$ is sufficient to determine the direction of the tangent to the surface.

The distance of two points P and Q on a curve is found by integrating

$$ds^2 = \sum_{i=1}^{3} dx_i \, dx_i = d\mathbf{x} \cdot d\mathbf{x} \tag{2–5}$$

along the curve; and substituting for $d\mathbf{x}$ the values (2–3), we find

$$ds^2 = (\mathbf{x}_u \, du + \mathbf{x}_v \, dv) \cdot (\mathbf{x}_u \, du + \mathbf{x}_v \, dv) = E \, du^2 + 2F \, du \, dv + G \, dv^2, \tag{2–6}$$

where

$$E = \mathbf{x}_u \cdot \mathbf{x}_u, \qquad F = \mathbf{x}_u \cdot \mathbf{x}_v, \qquad G = \mathbf{x}_v \cdot \mathbf{x}_v. \tag{2–7}$$

The E, F, G are functions of u and v. The distance between P and Q on the curve $u = u(t), v = v(t)$ can now be expressed as follows:

$$s = \int_{t_0}^{t} \sqrt{E \left(\frac{du}{dt}\right)^2 + 2F \frac{du}{dt} \frac{dv}{dt} + G \left(\frac{dv}{dt}\right)^2} \, dt.$$

The expression (2–6) for ds^2 is called the *first fundamental form* of the surface. It is a quadratic differential form; its square root ds can be taken as the length $|d\mathbf{x}|$ of the vector differential $d\mathbf{x}$ on the surface, and is called the *element of arc*. Since ds is a length,

$$E\,du^2 + 2F\,du\,dv + G\,dv^2$$

is always positive (except zero for $du = dv = 0$), as long as we study real surfaces; such a form is called *positive definite*. Since

$$ds^2 = \frac{1}{E}\,(E\,du + F\,dv)^2 + \frac{EG - F^2}{E}\,dv^2$$

and

$$E = \mathbf{x}_u \cdot \mathbf{x}_u > 0,$$

we see that

$$EG - F^2 > 0.$$

This also follows from Eq. (1–10), because $\mathbf{x}_u \times \mathbf{x}_v \neq 0$, so that

$$(\mathbf{x}_u \times \mathbf{x}_v) \cdot (\mathbf{x}_u \times \mathbf{x}_v) = (\mathbf{x}_u \cdot \mathbf{x}_u)(\mathbf{x}_v \cdot \mathbf{x}_v) - (\mathbf{x}_u \cdot \mathbf{x}_v)^2 = EG - F^2. \quad (2\text{–}8)$$

This inequality continues to hold under a change of curvilinear coordinates, since according to Eq. (1–13) for the new $\overline{E}$, $\overline{F}$, $\overline{G}$ the equation holds

$$\overline{E}\,\overline{G} - \overline{F}^2 = \begin{pmatrix} u & v \\ \bar{u} & \bar{v} \end{pmatrix}^2 (EG - F^2). \quad (2\text{–}8\text{a})$$

With the aid of E, F, G we can express the angle α of two tangent directions to the surface given by du/dv, $\delta u/\delta v$. Then

$$d\mathbf{x} = \mathbf{x}_u\,du + \mathbf{x}_v\,dv, \qquad \delta\mathbf{x} = \mathbf{x}_u\,\delta u + \mathbf{x}_v\,\delta v$$

and

$$\cos\alpha = \frac{d\mathbf{x} \cdot \delta\mathbf{x}}{|d\mathbf{x}||\delta\mathbf{x}|} = \frac{\mathbf{x}_u \cdot \mathbf{x}_u\,du\,\delta u + (\mathbf{x}_u \cdot \mathbf{x}_v)(du\,\delta v + dv\,\delta u) + (\mathbf{x}_v \cdot \mathbf{x}_v)\,dv\,\delta v}{|d\mathbf{x}||\delta\mathbf{x}|}$$

$$= \frac{E\,du\,\delta u + F(du\,\delta v + dv\,\delta u) + G\,dv\,\delta v}{\sqrt{E\,du^2 + 2F\,du\,dv + G\,dv^2}\,\sqrt{E\,\delta u^2 + 2F\,\delta u\,\delta v + G\,\delta v^2}}$$

$$= E\,\frac{du}{ds}\,\frac{\delta u}{\delta s} + F\left(\frac{du}{ds}\,\frac{\delta v}{\delta s} + \frac{dv}{ds}\,\frac{\delta u}{\delta s}\right) + G\,\frac{dv}{ds}\,\frac{\delta v}{\delta s}. \quad (2\text{–}9)$$

The following cases of Eq. (2–9) are particularly important:

1. When $\alpha = \pi/2$ we obtain the condition of orthogonality of two directions on the surface:

$$E\,du\,\delta u + F(du\,\delta v + dv\,\delta u) + G\,dv\,\delta v = 0. \quad (2\text{–}10)$$

2. The angle θ of the parametric lines $u = $ constant (hence $du = 0$, dv arbitrary) and $v = $ constant (hence δu arbitrary, $\delta v = 0$), is given by

$$\cos \theta = \frac{F \, dv \, \delta u}{\sqrt{G \, dv^2} \sqrt{E \, \delta u^2}} = \frac{F}{\sqrt{EG}},$$

$$\sin \theta = \frac{\sqrt{EG - F^2}}{\sqrt{EG}}. \tag{2-11}$$

3. *The parametric curves are orthogonal if $F = 0$.*

EXAMPLES. (1) *Sphere* (coordinates in Eqs. (1–7)). Squaring of the elements of the first row in M gives $E = a^2$, and similarly $F = 0$, $G = a^2 \cos^2 \theta$:

$$ds^2 = a^2 \, d\theta^2 + a^2 \cos^2 \theta \, d\varphi^2. \tag{2-12}$$

Since $F = 0$, meridians and parallels are shown to be orthogonal.

The *loxodromes* on the sphere (see Exercise 19, Section 1–11) are the curves which intersect the planes through a diameter, and hence the meridians, at a given angle α. Let the meridians be the curves $\varphi = $ const. Then, in Eq. (2–9), $dv = d\varphi = 0$, and therefore

$$\cos \alpha = \frac{E \, du \, \delta u}{ds \sqrt{E \, \delta u^2}} = a \frac{du}{ds} = a \frac{d\theta}{ds},$$

$$\cos^2 \alpha \, (d\theta^2 + \cos^2 \theta \, d\varphi^2) = d\theta^2,$$

$$d\theta / \cos \theta = \pm \cot \alpha \, d\varphi,$$

$$\pm \, (\varphi + c) \cot \alpha = \ln \tan \left(\frac{\theta}{2} + \frac{\pi}{4} \right).$$

The $\pm$ indicates that the loxodromes can wind around the sphere in a right-handed or a left-handed sense. When for given c and α (and the $+$ sign) the angle increases from $-\infty$ via $-c$ to $+\infty$, $\tan \left(\frac{\theta}{2} + \frac{\pi}{4} \right)$ increases from 0 via 1 to ∞, and θ from $-\frac{\pi}{2}$ via 0 to $\frac{\pi}{2}$. This shows that the loxodromes wind in spiral-like fashion around the poles as asymptotic points. Loxodromes on the sphere (and in particular on the globe) are also called *rhumb lines*.

(2) *Surface of revolution.* When the Z-axis is taken as the axis of revolution of the curve $z = f(x)$ in the plane $y = 0$ (this curve is the *profile* of the surface), the resulting surface (Fig. 2–4) can be given by the equations

$$x = r \cos \varphi, \qquad y = r \sin \varphi, \qquad z = f(r). \tag{2-13}$$

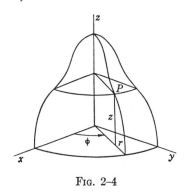

Fig. 2–4

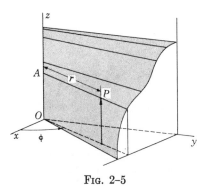

Fig. 2–5

The curves $r = $ constant are the parallels, and the curves $\varphi = $ constant are the meridians of the surface.   Here

$$M = \left\| \begin{array}{ccc} \cos\varphi & \sin\varphi & f'(r) \\ -r\sin\varphi & r\cos\varphi & 0 \end{array} \right\|$$

and

$$ds^2 = (1 + f'^2)\, dr^2 + r^2\, d\varphi^2. \qquad (2\text{–}14)$$

When

$$r = u \sin\alpha, \qquad f(r) = r \cot\alpha = u\cos\alpha \qquad (\alpha = \text{constant}), \quad (2\text{–}15)$$

we have a cone of revolution (see Eq. (1–8)) with the first fundamental form:

$$ds^2 = \csc^2\alpha\, dr^2 + r^2\, d\varphi^2 = du^2 + u^2 \sin^2\alpha\, d\varphi^2. \qquad (2\text{–}16)$$

Meridians and parallels are again orthogonal.

(3) *Right conoid.*   Such a surface is generated by a straight line moving parallel to a plane (here the plane $z = 0$) and intersecting a line perpendicular to this plane (here the Z-axis, Fig. 2–5).   Then its equation is

$$x = r\cos\varphi, \qquad y = r\sin\varphi, \qquad z = f(\varphi). \qquad (2\text{–}17)$$

The curves $r = $ constant are the loci of the points at fixed distance r from the Z-axis, and the curves $\varphi = $ constant are straight lines.   Here

$$M = \left\| \begin{array}{ccc} \cos\varphi & \sin\varphi & 0 \\ -r\sin\varphi & r\cos\varphi & f' \end{array} \right\|$$

and

$$ds^2 = dr^2 + (r^2 + f'^2)\, d\varphi^2. \qquad (2\text{–}18)$$

The coordinate curves $r = $ constant and $\varphi = $ constant are orthogonal.

When $f(\varphi) = a\varphi + b$ $(a, b$ constants) we have the *right helicoid* (see Section 1–3), for which

$$ds^2 = dr^2 + (r^2 + a^2)\, d\varphi^2. \tag{2–19}$$

The curves $r = $ constant are circular helices of equal pitch. Surfaces obtained by the motion of a straight line in space are called *ruled surfaces;* the right conoids are special cases of ruled surfaces, as are cones and cylinders.

2–3 Normal, tangent plane. All vectors $d\mathbf{x}/dt$ through P tangent to the surface satisfy the Eq. (2–2) and therefore lie in the plane of the vectors $\mathbf{x}_u$ and $\mathbf{x}_v$, uniquely determined at all points where $\mathbf{x}_u \times \mathbf{x}_v \neq 0$ (compare with Eq. (1–10)). This plane is the *tangent plane* at P to the surface. Its equation is

$$\mathbf{X} = \mathbf{x} + \lambda\mathbf{x}_u + \mu\mathbf{x}_v, \qquad \lambda, \mu \text{ parameters}, \tag{3–1a}$$

or

$$\begin{vmatrix} X_1 - x_1 & X_2 - x_2 & X_3 - x_3 \\ \dfrac{\partial x_1}{\partial u} & \dfrac{\partial x_2}{\partial u} & \dfrac{\partial x_3}{\partial u} \\ \dfrac{\partial x_1}{\partial v} & \dfrac{\partial x_2}{\partial v} & \dfrac{\partial x_3}{\partial v} \end{vmatrix} = 0; \tag{3–1b}$$

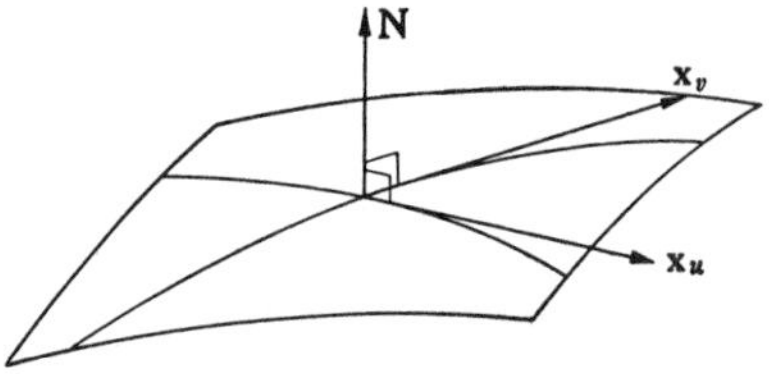

Fig. 2–6

where the derivatives are taken at P (x_1, x_2, x_3).

The surface normal, *normal* for short, is the line at P perpendicular to the tangent plane. As unit vector in this normal we take (Fig. 2–6):

$$\mathbf{N} = \frac{\mathbf{x}_u \times \mathbf{x}_v}{|\mathbf{x}_u \times \mathbf{x}_v|} = \frac{\mathbf{x}_u \times \mathbf{x}_v}{\sqrt{EG - F^2}}. \tag{3–2}*$$

We also might have taken $-\mathbf{N}$ as unit normal vector, since the sense of $\mathbf{N}$ depends on the labeling of the coordinate curves.

Since

$$\mathbf{x}_u \cdot \mathbf{N} = 0, \qquad \mathbf{x}_v \cdot \mathbf{N} = 0, \tag{3–3}$$

we conclude from Eq. (1–2), using the theory of contact presented in Section 1–7:

The tangent plane has, of all planes through P, the highest contact with the surface. This contact is (at least) of order one.

---

* $\sqrt{EG - F^2}$ will always mean $+\sqrt{EG - F^2}$.

In the texts on the calculus * it is shown that the area of a region R on the surface is given by

$$A = \iint \sqrt{EG - F^2}\, du\, dv.$$

The formula can be made plausible by the consideration that the area of a small parallelogram bounded by the two vectors $d\mathbf{x}$ and $\delta\mathbf{x}$ is $dA = |d\mathbf{x} \times \delta\mathbf{x}| = \sqrt{EG - F^2}\, |du\, \delta v - \delta u\, dv|$. Taking for $d\mathbf{x}$ and $\delta\mathbf{x}$ vectors tangent to the coordinate lines ($\delta u = 0$, $dv = 0$), and writing dv for δv (in accordance with the established custom of integration theory), we obtain

$$dA = \sqrt{EG - F^2}\, du\, dv. \tag{3–4}$$

EXAMPLES. (1) *Circular cone.* Here, according to Eq. (1–8) and Eq. (2–14) with

$$f(r) = r \cot \alpha, \qquad r = u \sin \alpha,$$

we have

$$\sqrt{EG - F^2} = u \sin \alpha,$$

and the coordinates of **N** are

$$\mathbf{N}(N_1 = -\cos \alpha \cos \varphi, \qquad N_2 = -\cos \alpha \sin \varphi, \qquad N_3 = \sin \alpha).$$

N is independent of r (or u), which means that at all points along a generating line the tangent plane is the same.

(2) *Right helicoid.* According to Eq. (2–19):

$$\sqrt{EG - F^2} = \sqrt{r^2 + a^2},$$

and according to Eq. (2–18):

$$\mathbf{N}\left(\frac{a \sin \varphi}{\sqrt{r^2 + a^2}}, \quad \frac{-a \cos \varphi}{\sqrt{r^2 + a^2}}, \quad \frac{r}{\sqrt{r^2 + a^2}}\right).$$

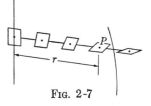

FIG. 2-7

If we call γ the angle that **N** makes with the Z-axis, we have

$$\cos \gamma = N_3, \qquad \cot \gamma = \frac{r}{a}.$$

This shows that when a point P moves along a generating line the tangent plane turns about this line such that its angle with the central axis changes from $0°$ on this axis to $90°$, the tangent of this angle changing proportionally to r (Fig. 2–7). This will be more explicitly discussed in Section 5–5.

---

* e.g. P. Franklin, loc. cit., Section 1–1.

(3) *Tangential developable.* This is the surface generated by the tangent lines to a space curve. If this curve C is given by the equation

$$\mathbf{x} = \mathbf{x}(s), \text{ with unit tangent vector } \mathbf{t} = \mathbf{t}(s),$$

then the equation of the surface is (Fig. 2–8)

$$\mathbf{y}(s, v) = \mathbf{x}(s) + v\mathbf{t}(s), \tag{3-5}$$

where v is the distance from a point P on a generating line of the surface to its tangent point A on C. Here

$$\mathbf{y}_s = \mathbf{t} + v\kappa\mathbf{n}, \qquad \mathbf{y}_v = \mathbf{t}, \qquad \mathbf{y}_s \times \mathbf{y}_v = v\kappa\mathbf{n} \times \mathbf{t} = -v\kappa\mathbf{b},$$
$$\mathbf{N} = \mathbf{b}(\text{or } -\mathbf{b}).$$

The tangent plane along a generator coincides with the osculating plane of the point on C through which the generator passes. It is therefore the same along a generator.

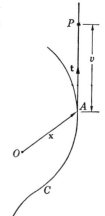

Fig. 2–8

EXERCISES

1. The following surfaces are given in parametric form.

(a) *Ellipsoid:*

$$x = a \sin u \cos v, \qquad y = b \sin u \sin v, \qquad z = c \cos u.$$

(b) *Hyperboloid of two sheets:*

$$x = a \sinh u \cos v, \qquad y = b \sinh u \sin v, \qquad z = c \cosh u.$$

(c) *Cone:*

$$x = a \sinh u \sinh v, \qquad y = b \sinh u \cosh v, \qquad z = c \sinh u.$$

(d) *Elliptic paraboloid:*

$$x = au \cos v, \qquad y = bu \sin v, \qquad z = u^2.$$

(e) *Hyperbolic paraboloid:*

$$x = au \cosh v, \qquad y = bu \sinh v, \qquad z = u^2.$$

Find the equations of these surfaces in the form $F(x, y, z) = 0$. What kind of curves are the coordinate curves $u = $ constant, $v = $ constant in each case?

2. Show that (a) the hyperbolic paraboloid can also be given by the equations:

$$x = a(u + v), \qquad y = b(u - v), \qquad z = uv,$$

and (b) the hyperboloid of one sheet is given by

$$x = a\frac{u - v}{u + v}, \qquad y = b\frac{1 + uv}{u + v}, \qquad z = c\frac{uv - 1}{u + v}.$$

What are the curves $u = $ constant, $v = $ constant?

(c) Also study the surface $x = u$, $y = v$, $z = uv$.

3. Given a surface by the equation $z = f(x, y)$.   (a) Find the first fundamental form and **N**.   (b) Do the same if the equation is $F(x, y, z) = 0$.

4. Show that the element of area of the surface with equation $z = f(x, y)$ is $dA = \sqrt{1 + p^2 + q^2}\, dx\, dy$, $p = \partial z/\partial x$, $q = \partial z/\partial y$.

5. Show that the orthogonal trajectories of the family of curves given by

$$M\, du + N\, dv = 0$$

are given by

$$(EN - FM)\, du + (FN - GM)\, dv = 0,$$

and use this formula to find the orthogonal trajectories of the circles $r = a \cos \theta$ in the plane for all values of a (use polar coordinates).

6. Show that the necessary and sufficient condition that the curves

$$A\, du^2 + 2B\, du\, dv + C\, dv^2 = 0, \qquad A, B, C \text{ functions of } u, v,$$

form an orthogonal net is

$$EC - 2FB + GA = 0.$$

7. Show that the curves $dr^2 - (r^2 + a^2)\, d\varphi^2 = 0$ on the right helicoid (see Eq. (2–19)) form an orthogonal net.

8. When the first fundamental form of a surface can be written in the form

$$ds^2 = du^2 + G(u, v)\, dv^2,$$

the curves $u = $ constant cut equal segments from all curves $v = $ constant.   Since the curves $u = $ constant, $v = $ constant are also orthogonal, we call the curves $u = $ constant *parallel*.

9. Assume the parametric curves to be orthogonal.   Show that the differential equation of the curves bisecting the angles of the parametric curves is

$$E\, du^2 - G\, dv^2 = 0.$$

10. Show that the curves on the cone (1–8) given by

$$u = c \exp(\varphi \sin \alpha \cot \beta),$$

cut the generating lines at constant angle β.   Show that they project on the XOY-plane as logarithmic spirals.

11. What is the length of a loxodrome on a sphere which starts at the equator at an angle α with the meridian and ends up by winding around a pole?

12. The locus of the mid-points of the chords of a circular helix is a right helicoid.

13. Show that the locus of the projections of the center of an ellipsoid

$$\frac{x^2}{a^2} + \frac{y^2}{b^2} + \frac{z^2}{c^2} = 1$$

on its tangent planes has the equation

$$(x^2 + y^2 + z^2)^2 = a^2x^2 + b^2y^2 + c^2z^2.$$

This is the *pedal surface* (compare with Section 1–13, Exercise 14) of the ellipsoid with respect to its center.   It is called *Fresnel's elasticity surface*.

2–4 Developable surfaces. Tangent developables share with cones and cylinders the property that they have a constant tangent plane along a generating line. Their tangent planes therefore depend on only one parameter. Such surfaces can be considered as the *envelope of* a one-parameter family of planes. We shall show that they are the only surfaces that can be considered as the envelope of such a family of ∞^1 planes.

Such a family can be given by the equation

$$\mathbf{x} \cdot \mathbf{a} + p = 0, \tag{4–1}$$

where the $\mathbf{a}$ and p depend on a parameter u. We exclude the case that

$$\mathbf{a}' = d\mathbf{a}/du = 0,$$

which gives a family of parallel planes, as does the case that $\mathbf{a}$ and $\mathbf{a}'$ are collinear. The planes determined by the parameters $u = u_1$ and $u = u_2(u_1 < u_2)$ then intersect in a straight line, which also lies in the plane

$$\mathbf{x} \cdot \{\mathbf{a}(u_1) - \mathbf{a}(u_2)\} + p(u_1) - p(u_2) = 0,$$

or, applying Rolle's theorem,

$$x_1 a_1'(v_1) + x_2 a_2'(v_2) + x_3 a_3'(v_3) + p'(w_1) = 0, \qquad u_1 \leqslant v_i \leqslant u_2, \qquad u_1 \leqslant w_1 \leqslant u_2.$$

When $u_2 \to u_1$ we find that this line takes a limiting position given by

$$\mathbf{x} \cdot \mathbf{a} + p = 0, \qquad \mathbf{x} \cdot \mathbf{a}' + p' = 0, \qquad p' = dp/du. \tag{4–2}$$

This line is called the *characteristic* of the plane (4–1).

The planes determined by the parameters $u = u_1$, $u = u_2$, and $u = u_3(u_1 < u_2 < u_3)$ intersect in a point, which also lies in the planes

$$\mathbf{x} \cdot \mathbf{a}'(v_1) + p'(w_1) = 0, \quad \mathbf{x} \cdot \mathbf{a}'(v_2) + p'(w_2) = 0, \quad \mathbf{x} \cdot \mathbf{a}''(v_3) + p''(w_3) = 0;$$
$$u_1 \leqslant v_1 \leqslant u_2, \quad u_2 \leqslant v_2 \leqslant u_3, \quad v_1 \leqslant v_3 \leqslant v_2, \quad u_1 \leqslant w_1 \leqslant u_2, \quad u_2 \leqslant w_2 \leqslant u_3, \quad w_1 \leqslant w_3 \leqslant w_2.$$

When $u_3 \to u_2 \to u_1$ we find that this point takes a limiting position given by

$$\mathbf{x} \cdot \mathbf{a} + p = 0, \qquad \mathbf{x} \cdot \mathbf{a}' + p' = 0, \qquad \mathbf{x} \cdot \mathbf{a}'' + p'' = 0. \tag{4–3}$$

This point is called the *characteristic point* of the plane (4–1). It lies on the characteristic line. It does not exist when $(\mathbf{a}\,\mathbf{a}'\,\mathbf{a}'') = 0$, in which case the vector field $\mathbf{a}(u)$ is plane. But the vector $\mathbf{a}$ is perpendicular to plane (4–1). When the vectors $\mathbf{a}(u)$ are parallel to a plane π, the planes (4–1) are all parallel to the direction perpendicular to π. In this case the envelope of the planes (4–1) is therefore a *cylinder* with generating lines perpendicular to the plane π (except in the case that the planes form a pencil and the "envelope" is a straight line).

When the characteristic point is the same for all planes the envelope is a *cone* generated by all characteristic lines.

In the general case there exists a locus of characteristic points, which is a curve C. We shall show that the characteristic line is tangent to C at its characteristic point. For this purpose let us consider Eq. (4–3) solved for $\mathbf{x}$, which then becomes a function of u. Differentiation of the first two equations of (4–3) shows that

$$\mathbf{x}' \cdot \mathbf{a} + \mathbf{x} \cdot \mathbf{a}' + p' = 0, \qquad \mathbf{x}' \cdot \mathbf{a}' + \mathbf{x} \cdot \mathbf{a}'' + p'' = 0,$$

which, when compared with Eq. (4–3), gives

$$\mathbf{x}' \cdot \mathbf{a} = 0, \qquad \mathbf{x}' \cdot \mathbf{a}' = 0. \tag{4–4}$$

This means that the tangent vector $\mathbf{x}'$ to C has the direction of line (4–2). Since it passes through the characteristic point, $\mathbf{x}'$ must lie in the characteristic line. Similarly, differentiating the first equation of (4–4), we obtain

$$\mathbf{x}' \cdot \mathbf{a}' + \mathbf{x}'' \cdot \mathbf{a} = 0.$$

Comparing this with the second equation of (4–4) we find that $\mathbf{x}'' \cdot \mathbf{a} = 0$, which with $\mathbf{x}' \cdot \mathbf{a} = 0$ means that the vector $\mathbf{x}' \times \mathbf{x}''$ is parallel to $\mathbf{a}$. The osculating plane of the curve C at the characteristic point is therefore identical with the plane (4–1). We have thus found the following theorem.

A family of ∞^1 planes, which are not all parallel and which do not form a pencil, has as its envelope either a cylinder, a cone, or a tangential developable. This envelope is generated by the characteristic lines of the planes, which in the case of a cone all pass through the one characteristic point and in the case of a tangential developable are all tangent to the locus of the characteristic points.

The locus of the characteristic points is called the *edge of regression* (French: *arête de rebroussement;* German: *Rückkehrkante*). It reduces to a point in the case of a cone. Its name is due to the fact that *the intersection of a tangential developable with the normal plane to this edge of regression at one of its points P has a cusp at that point.*

This property expresses the fact that the developable consists of two sheets which are tangent at the edge of regression along a sharp edge. We show this property analytically by taking the trihedron $(\mathbf{t}, \mathbf{n}, \mathbf{b})$ at P as the coordinate tetrahedron (y_1, y_2, y_3) to which the surface is referred:

$$\mathbf{y} = \mathbf{x} + u\mathbf{t} = s\mathbf{t} + \frac{s^2}{2}\kappa\mathbf{n} + \frac{s^3}{6}\left(-\kappa^2\mathbf{t} + \kappa'\mathbf{n} + \kappa\tau\mathbf{b}\right) + o(s^3) \tag{4–5}$$

$$+ u\left[\mathbf{t} + s\kappa\mathbf{n} + \frac{s^2}{2}\left(-\kappa^2\mathbf{t} + \kappa'\mathbf{n} + \kappa\tau\mathbf{b}\right) + o(s^2)\right]\cdot$$

The intersection with the normal plane at P is determined by $y_1 = 0$, or by

$$s - \frac{\kappa^2}{6} s^3 + o(s^3) + u\left[1 - \frac{\kappa^2}{2} s^2 + o(s^2)\right] = 0,$$

hence

$$u = -s - \frac{\kappa^2}{3} s^3 + o(s^3).$$

Substituting this value of u into the expressions for y_2 and y_3, we obtain

$$y_2 = \frac{\kappa}{2} s^2 + o(s^2) + u\{\kappa s + o(s)\} = -\frac{\kappa}{2} s^2 + o(s^2),$$

$$y_3 = \frac{\kappa\tau}{6} s^3 + o(s^3) + u\left\{\frac{\kappa\tau}{2} s^2 + o(s^2)\right\} = -\frac{\kappa\tau}{3} s^3 + o(s^3).$$

We thus obtain for the required intersection in first approximation

$$y_3^2 = -\tfrac{8}{9}\tau^2 R y_2^3, \tag{4-6}$$

which shows that the intersection of the surface with the normal plane to the curve has a cusp with the principal normal as tangent (Fig. 2–9). The edge of regression appears as a sharp edge on the surface (Fig. 2–10). For this reason it is sometimes called the *cuspidal edge*. Comparison of (4–6) with Chapter 1, Eq. (6–4) shows that the cusp of this normal section has a sense opposite to that of the projection of the cuspidal edge on the normal plane at P.

Cylinders, cones, and tangential developables are called *developable surfaces*, or simply *developables* (German, sometimes: *Torsen*). If we let one

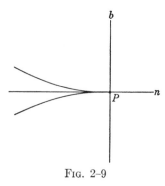

Fig. 2–9

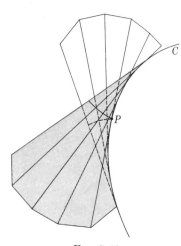

Fig. 2–10

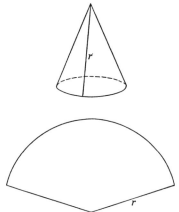

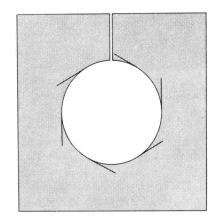

Fig. 2–11

Fig. 2–12

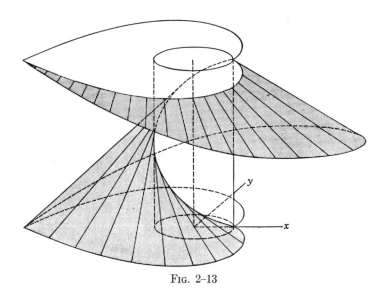

Fig. 2–13

tangent plane coincide with a plane π, we can turn the "next" tangent plane about the characteristic line into π, then the "next" plane, etc., and in this way develop the surface into the plane. Conversely, we can obtain a developable surface by bending a plane, without stretching or shrinking, so that the metrical properties of the plane are unchanged. This can be demonstrated with a piece of paper, which can be bent without stretching around a cylinder. A piece of paper in the form of a circular sector can be bent around a cone; the radius of the sector becomes the slant height of the cone (Fig. 2–11). A tangent developable of a circular helix can be obtained by cutting a circle out of a piece of paper and twisting the remaining part of the paper along an appropriate circular helix on a cylinder (Fig. 2–12); the tangents to the circle become the generating lines of the developable (Fig. 2–13). We shall later give an exact proof of this property of a developable.

The two sheets of the developable are characterized by $v > 0$ and $v < 0$ in Eq. (3–5); $v = 0$ gives the edge of regression.

EXAMPLES. (1) *Developable helicoid* (Fig. 2–13). This is the surface generated by the tangents to a circular helix. Since

$$x_1 = a \cos u, \qquad x_2 = a \sin u, \qquad x_3 = bu,$$

the surface has the equations:

$$y_1 = a \cos u - av \sin u, \quad y_2 = a \sin u + av \cos u, \quad y_3 = bu + bv.$$

The parameters are u and v. This surface intersects the XOY-plane in the curve for which $y_3 = 0$ or $v = -u$:

$$y_1 = a \cos u + au \sin u,$$
$$y_2 = a \sin u - au \cos u,$$

which is a circle involute. This also follows from the fact that the projection of the tangent to the helix between the point P on the helix and the intersection with the XOY-plane is $bu \tan \gamma = au$. (Compare with Sections 1–2 and 1–11.)

The circle involute intersects itself in an infinity of points (only one of these points is constructed in Fig. 2–14). As the tangent surface participates in the helicoidal motion by which the circular helix is generated, we

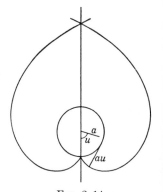

FIG. 2–14

see that *the developable helicoid has an infinity of double lines* where the surface intersects itself.   These double lines are also helices.

(2) *Polar developable.*   This is the surface enveloped by the normal planes of a space curve, the equations of which have the form

$$(\mathbf{X} - \mathbf{x}) \cdot \mathbf{t} = 0, \qquad (4\text{--}7)$$

where $\mathbf{x}$ and $\mathbf{t}$ are functions of the arc length s of the curve, and $\mathbf{X}$ is a generic point, independent of s.   The parameter is s.   The characteristics are given by Eq. (4–7) and

$$-\mathbf{t} \cdot \mathbf{t} + (\mathbf{X} - \mathbf{x}) \cdot \kappa\mathbf{n} = 0, \qquad (4\text{--}8\text{a})$$

or

$$(\mathbf{X} - \mathbf{x}) \cdot \mathbf{n} = \kappa^{-1} = R; \qquad (4\text{--}8)$$

the characteristic points are given by Eqs. (4–7), (4–8), and

$$-\mathbf{t} \cdot \mathbf{n} + (\mathbf{X} - \mathbf{x}) \cdot (-\kappa\mathbf{t} + \tau\mathbf{b}) = R',$$

obtained by differentiating either Eq. (4–8) or (4–8a).   Hence

$$(\mathbf{X} - \mathbf{x}) \cdot \mathbf{b} = TR'. \qquad (4\text{--}9)$$

The edge of regression of the polar developable is the curve

$$\mathbf{X} = \mathbf{x} + R\mathbf{n} + TR'\mathbf{b},$$

which according to Chapter 1, Eq. (7–5) is the locus of the centers of the osculating spheres; the generating lines of the developable are the polar axes.   This polar developable can be taken as the space analog of the evolute of a plane curve, which is the envelope of the normal lines.   This can be seen by interpreting Eqs. (4–7) and (4–8) for plane curves.   We can also say that just as two consecutive normals of a plane curve intersect in a point of the evolute, so do three consecutive normal planes of a space curve intersect in a point of the locus of the centers of the osculating spheres.

(3) *Tangential developable.*   The envelope of the osculating planes of a space curve is the tangential developable, since

$$(\mathbf{X} - \mathbf{x}) \cdot \mathbf{b} = 0,$$
$$-\mathbf{t} \cdot \mathbf{b} - (\mathbf{X} - \mathbf{x}) \cdot \tau\mathbf{n} = 0, \quad \text{or} \quad (\mathbf{X} - \mathbf{x}) \cdot \mathbf{n} = 0,$$
$$-\mathbf{t} \cdot \mathbf{n} + (\mathbf{X} - \mathbf{x}) \cdot (-\kappa\mathbf{t} + \tau\mathbf{b}) = 0, \quad \text{or} \quad (\mathbf{X} - \mathbf{x}) \cdot \mathbf{t} = 0$$

gives the curve itself as the edge of regression and its tangents as the characteristic lines.

(4) *Rectifying developable*. The envelope of the rectifying planes of a space curve is determined by

$$(\mathbf{X} - \mathbf{x}) \cdot \mathbf{n} = 0, \qquad (\mathbf{X} - \mathbf{x}) \cdot (-\kappa\mathbf{t} + \tau\mathbf{b}) = 0$$

as the locus of lines in the rectifying plane passing through the point P of the curve and making an angle $\tan^{-1}\frac{\kappa}{\tau}$ with the positive tangent direction. We shall meet this developable later, where the name will be explained.

A space curve is the locus of ∞^1 points, a developable the envelope of ∞^1 planes. This leads to a duality between space curves and developables which can be indicated as follows:

Curves	*Developables*
2 points determine a line.	2 planes determine a line.
3 points determine a plane.	3 planes determine a point.
2 consecutive points on a curve determine a tangent line.	2 consecutive planes of the family of ∞^1 planes determine a characteristic line.
3 consecutive points on a curve determine an osculating plane.	3 consecutive planes of the family of ∞^1 planes determine a characteristic point.
The curve is the envelope of ∞^1 tangents.	The developable is generated by ∞^1 characteristic lines.
The curve is the edge of regression of the developable enveloped by the osculating planes.	The developable is the envelope of the osculating planes of the curve **generated** by the characteristic points.
Plane curve.	Cone.

The analogy is not complete. Two points, for instance, always determine a straight line, but two planes may be parallel. To extend the analogy to cases which involve elements at infinity and metrical relationships involves a deeper study of affine and non-Euclidean geometry in terms of projective geometry.

Developable surfaces are generated by straight lines, but not all surfaces generated by straight lines, the so-called *ruled surfaces*, are developable. An example is the hyperboloid of one sheet, which is a ruled surface, but not developable, since the tangent plane varies when the point of tangency moves along a generating line. We discuss this further in Section 5–5.

1. Find the envelope of the planes

(a) $\qquad u^3 - 3u^2 x_1 + 3u x_2 - x_3 = 0.$

(b) $\quad x_1 \sin u - x_2 \cos u + x_3 \tan \theta - au = 0.$ (θ, a are constants)

2. Show that the tangent developable of the cubic parabola

$$x_1 = u, \qquad x_2 = u^2, \qquad x_3 = u^3$$

is the surface

$$4(x_2 - x_1^2)(x_1 x_3 - x_2^2) - (x_1 x_2 - x_3)^2 = 0.$$

3. The rectifying developable of a curve is the polar developable of all its involutes.

4. When the polar developable of a curve is a cone the curve lies on a sphere.

5. When the envelope of the rectifying planes of a curve is a cone the curve satisfies the condition $\tau/\kappa = as + b$, where a and b are constants. Can the envelope of the osculating planes be a cone?

6. Show that the envelope of the planes which form with the three coordinate planes a tetrahedron of constant volume, is the surface $x\,y\,z =$ constant.

7. The osculating plane of the edge of regression at a point P of a curve cuts the developable in a generating line (counted twice) and in a curve of which the curvature at P is $\frac{3}{4}\kappa$, κ being the curvature of the edge of regression at P.

8. The rectifying plane of the edge of regression cuts the developable in a generating line (counted once) and in a curve with a point of inflection at P, the generating line being the inflectional tangent.

9. *Envelopes of straight lines in the plane.* If a family of ∞^1 lines in the plane is given by $\mathbf{a} \cdot \mathbf{x} + p = 0$ (Eq. (4-1), Section 2-4), where $\mathbf{a} = \mathbf{a}(s)$, $p = p(s)$ then its envelope, if it exists, is given by

$$\mathbf{a} \cdot \mathbf{x} + p = 0, \qquad \mathbf{a}' \cdot \mathbf{x} + p' = 0.$$

Prove that the lines of the family are tangent to the envelope. When is there no envelope?

10. Find the envelope of a line of constant length a which moves with its ends on two perpendicular lines. This curve is a *hypocycloid of four cusps.*

11. Have the principal normals of a curve an envelope?

12. Have the binormals of a curve an envelope?

2-5 Second fundamental form. Meusnier's theorem.

The geometry of surfaces depends on two quadratic differential forms. We have already introduced the first of them, which represents ds^2. The *second fundamental form* can be obtained by taking on the surface a curve C passing through a point P, and considering the curvature vector of C at P (see Chapter 1, Eq. (4-4)). When $\mathbf{t}$ is the unit tangent vector of C, this curvature vector $\mathbf{k}$

is equal to dt/ds. We now decompose $\mathbf{k}$ into a component $\mathbf{k}_n$ normal and a component $\mathbf{k}_g$ tangential to the surface (Fig. 2–15):

$$dt/ds = \mathbf{k} = \mathbf{k}_n + \mathbf{k}_g. \tag{5–1}$$

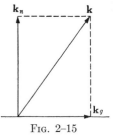

Fig. 2–15

The vector $\mathbf{k}_n$ is called the *normal curvature vector* and can be expressed in terms of the unit surface normal vector $\mathbf{N}$:

$$\mathbf{k}_n = \kappa_n \mathbf{N}, \tag{5–2}$$

where κ_n is the *normal curvature*. The vector $\mathbf{k}_n$ is determined by C alone (not by any choice of the sense of $\mathbf{t}$ or $\mathbf{N}$), the scalar κ_n depends for its sign on the sense of $\mathbf{N}$. The vector $\mathbf{k}_g$ is called the *tangential curvature vector* or *geodesic curvature vector*. We shall deal in this chapter with the properties of $\mathbf{k}_n$; those of $\mathbf{k}_g$ belong to the subject matter of Chapter 4.

From the equation $\mathbf{N} \cdot \mathbf{t} = 0$ we obtain by differentiation along C:

$$\frac{dt}{ds} \cdot \mathbf{N} = -\mathbf{t} \cdot \frac{d\mathbf{N}}{ds} = -\frac{d\mathbf{x}}{ds} \cdot \frac{d\mathbf{N}}{ds}, \tag{5–3}$$

or

$$\kappa_n = -\frac{d\mathbf{x} \cdot d\mathbf{N}}{d\mathbf{x} \cdot d\mathbf{x}}. \tag{5–4}$$

Let us study first the right-hand side of this equation.

Both $\mathbf{N}$ and $\mathbf{x}$ are surface functions of u and v (which in their turn depend on C). With the aid of the resulting identities

$$d\mathbf{N} = \mathbf{N}_u \, du + \mathbf{N}_v \, dv, \qquad d\mathbf{x} = \mathbf{x}_u \, du + \mathbf{x}_v \, dv, \tag{5–5}$$

we can write Eq. (5–4) in the form:

$$\kappa_n = -\frac{(\mathbf{x}_u \cdot \mathbf{N}_u) \, du^2 + (\mathbf{x}_u \cdot \mathbf{N}_v + \mathbf{x}_v \cdot \mathbf{N}_u) \, du \, dv + (\mathbf{x}_v \cdot \mathbf{N}_v) \, dv^2}{E \, du^2 + 2F \, du \, dv + G \, dv^2},$$

or

$$\kappa_n = \frac{e \, du^2 + 2f \, du \, dv + g \, dv^2}{E \, du^2 + 2F \, du \, dv + G \, dv^2}. \tag{5–6}$$

In this equation

$$e = -\mathbf{x}_u \cdot \mathbf{N}_u, \qquad 2f = -(\mathbf{x}_u \cdot \mathbf{N}_v + \mathbf{x}_v \cdot \mathbf{N}_u), \qquad g = -\mathbf{x}_v \cdot \mathbf{N}_v \tag{5–7}$$

are functions of u and v, which depend on the second derivatives of the $\mathbf{x}$ with respect to u and v, in this respect differing from E, F, G, which depend on first derivatives only. We write the denominator and numerator of Eq. (5–6) in the following form:

$$\boxed{\begin{aligned} \text{I} &= E\,du^2 + 2F\,du\,dv + G\,dv^2 = d\mathbf{x}\cdot d\mathbf{x}, \\ \text{II} &= e\,du^2 + 2f\,du\,dv + g\,dv^2 = -d\mathbf{x}\cdot d\mathbf{N}. \end{aligned}} \tag{5–8}$$

I is the first fundamental form, II the *second fundamental form*. Since $\mathbf{x}_u \cdot \mathbf{N} = 0$ and $\mathbf{x}_v \cdot \mathbf{N} = 0$, we can also write for e, f, and g:

$$e = \mathbf{x}_{uu} \cdot \mathbf{N}, \qquad f = \mathbf{x}_{uv} \cdot \mathbf{N}, \qquad g = \mathbf{x}_{vv} \cdot \mathbf{N}, \tag{5–9,}$$

or, using the expression (3–2) for $\mathbf{N}$:

$$e = \frac{(\mathbf{x}_{uu}\mathbf{x}_u\mathbf{x}_v)}{\sqrt{EG - F^2}} = \frac{\begin{vmatrix} x_{uu} & y_{uu} & z_{uu} \\ x_u & y_u & z_u \\ x_v & y_v & z_v \end{vmatrix}}{\sqrt{EG - F^2}}. \tag{5–10a}$$

Similarly,

$$f = \frac{(\mathbf{x}_{uv}\mathbf{x}_u\mathbf{x}_v)}{\sqrt{EG - F^2}}, \qquad g = \frac{(\mathbf{x}_{vv}\mathbf{x}_u\mathbf{x}_v)}{\sqrt{EG - F^2}}. \tag{5–10b}$$

These formulas (5–10) allow ready computation of e, f, g when the equation of the surface is given.

Incidentally, we can derive from Eq. (3–3) that

$$\mathbf{x}_u \cdot \mathbf{N}_v = \mathbf{x}_v \cdot \mathbf{N}_u,$$

so that Eq. (5–7) can be rewritten in the simpler form

$$e = -\mathbf{x}_u \cdot \mathbf{N}_u, \quad f = -\mathbf{x}_u \cdot \mathbf{N}_v = -\mathbf{x}_v \cdot \mathbf{N}_u, \quad g = -\mathbf{x}_v \cdot \mathbf{N}_v. \tag{5–7a}$$

Returning now to Eq. (5–4) or, what is its equivalent, Eq. (5–6), we see that the right-hand side depends only on u, v, and dv/du. The coefficients e, f, g, E, F, G are constants at P, so that κ_n is fully determined, at P, by the direction dv/du. All curves through P tangent to the same direction have therefore the same normal curvature (if the sense of $\mathbf{N}$ is the same for all these curves). Expressed in vector language (Fig. 2–16):

All curves through P tangent to the same direction have the same normal curvature vector.

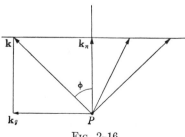

Fig. 2–16

When we momentarily give to $\mathbf{N}$ the sense of $\mathbf{k}_n$, and to $\mathbf{n}$ the sense of dt/ds, then we can express this theorem in the form obtained from Eq. (5–3):

$$\kappa \cos \varphi = \kappa_n, \qquad (5\text{–}11)$$

where $\varphi(0 \leqslant \varphi \leqslant \pi/2)$ is the angle between $\mathbf{N}$ and $\mathbf{n}$ (Fig. 2–16). This equation can be cast into another form for directions $\mathbf{t}$ for which $\kappa_n \neq 0$, hence also $\kappa \neq 0$. Such directions are called *nonasymptotic directions*. For curves in such directions we can write $R = \kappa^{-1}$, $R_n = \kappa_n^{-1}$. The quantities R and R_n are here positive, R_n represents the radius of curvature of a curve with tangent $\mathbf{t}$ and $\varphi = 0$. One such curve is the intersection of the surface with the plane at P through $\mathbf{t}$ and the surface normal; this curve is called the *normal section* of the surface at P in the direction of C. Eq. (5–11) now takes the form

$$\boxed{R_n \cos \varphi = R.} \qquad (5\text{–}12)$$

We can thus cast our previous theorem into the form:

The center of curvature C_1 of a curve C in a nonasymptotic direction at P is the projection on the principal normal of the center of curvature C_0 of the normal section which is tangent to C at P (Fig. 2–17). And in still other words:

If a set of planes be drawn through a tangent to a surface in a nonasymptotic direction, then the osculating circles of the intersections with the surface lie upon a sphere.

This theorem is known as *Meusnier's theorem*.

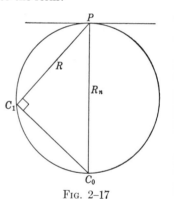

FIG. 2–17

Jean Baptiste Meusnier (1754–1793), a pupil of Monge at the school in Mézières, published the theorem in his *Mémoire sur la courbure des surfaces, Mémoires des savans étrangers* 10 (lu 1776), 1785, pp. 477–510, which he wrote after Monge had shown him Euler's paper (see Section 2–6). In 1783–1784, after Montgolfier's ascent in a balloon, he did fundamental research on "aerostation," and in this same period collaborated with Lavoisier in his work on the decomposition of water into its elements. He died as a revolutionary general during the siege of Mayence. See G. Darboux' account in *Eloges académiques* (Paris, 1912), pp. 218–262.

The last two versions of Meusnier's theorem do not hold for directions for which the second fundamental form II is zero, since in those directions, according to Eq. (5–6), $\kappa_n = 0$. We have seen that there are no (real) directions for which the first fundamental form I is zero. However, it may happen that there are real directions for which II is zero:

$$e\,du^2 + 2f\,du\,dv + g\,dv^2 = 0. \tag{5–13}$$

This happens, for instance, when there are straight lines on the surface. Directions satisfying Eq. (5–13) are called *asymptotic directions*. Curves having these directions are called *asymptotic curves* (German: *Haupttangentenkurven*). When Eq. (5–13) is satisfied, Eqs. (5–6) and (5–11) indicate that *a normal section in an asymptotic direction has a point of inflection.*

2–6 Euler's theorem. We shall now investigate the behavior of the normal curvature vector when the tangent direction at P varies. Its direction always remains that of the surface normal, but its length may vary for different directions. When the sense of **N** has been agreed upon, the sign of κ_n will express whether $\mathbf{k}_n$ has the sense of **N** or the opposite sense.

Since $I > 0$ the sign of κ depends only on II. There are three cases:

(1) II maintains the same sign whatever the direction may be. In this case II is a *definite* quadratic form; this is expressed by the condition that

$$eg - f^2 > 0.$$

(See the reasoning used for the case of I in Section 2–2.) The centers of curvature of the normal sections are all on the same side of the surface normal; the normal sections are all concave (or all convex). The point is called an *elliptic* point of the surface; an example is any point on the ellipsoid.

(2) II is a perfect square:

$$eg - f^2 = 0.$$

The surface behaves at the point like an elliptic point, except in one direction, where $\kappa_n = 0$; the curves in this direction have a point of inflection. The point is called *parabolic;* an example is any point on a cylinder.

(3) II does not maintain the same sign for all directions du/dv. In this case II is *indefinite;* the condition is

$$eg - f^2 < 0.$$

The normal sections are concave when they are cut out by planes in directions lying in one section of the tangent plane, convex when outside this

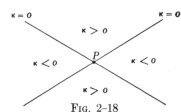

κ = 0   κ = 0

κ > 0

κ < 0   P   κ < 0

κ > 0

Fig. 2–18

section. The sections are separated by the directions in which $\kappa_n = 0$, the asymptotic directions (see Eq. (5–13) and Fig. 2–18). The point is called *hyperbolic (saddle point)*; an example is any point on a hyperbolic paraboloid (Fig. 2–19).

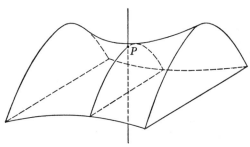

Fig. 2–19

The conditions $eg - f^2 \gtreqless 0$ do not depend on the choice of curvilinear coordinates on the surface. This is geometrically evident, since these formulas express geometrical properties of the surface. It follows analytically from Eq. (1–12) and

$$\mathbf{N}_{\bar{u}} = \mathbf{N}_u \frac{\partial u}{\partial \bar{u}} + \mathbf{N}_v \frac{\partial v}{\partial \bar{u}}, \qquad \mathbf{N}_{\bar{v}} = \mathbf{N}_u \frac{\partial u}{\partial \bar{v}} + \mathbf{N}_v \frac{\partial v}{\partial \bar{v}},$$

so that according to the definitions (5–7) and the resulting equation

$$eg - f^2 = (\mathbf{x}_u \times \mathbf{x}_v) \cdot (\mathbf{N}_u \times \mathbf{N}_v)$$

for the $\bar{e}, \bar{f}, \bar{g}$ defined in the new coordinate system, the equation holds (compare Eq. (1–13)):

$$\bar{e}\bar{g} - \bar{f}^2 = \begin{pmatrix} u & v \\ \bar{u} & \bar{v} \end{pmatrix}^2 (eg - f^2).$$

Now we take the equation of the surface in the form (1–2); $n = 3$ (omit the index 0 for the derivatives):

$$\mathbf{x} = \mathbf{x}_0 + h\mathbf{x}_u + k\mathbf{x}_v + \tfrac{1}{2}(h^2\mathbf{x}_{uu} + 2hk\mathbf{x}_{uv} + k^2\mathbf{x}_{vv}) \\ + \frac{1}{3!}\left(h\frac{\partial}{\partial u} + k\frac{\partial}{\partial v}\right)^3 \mathbf{x}(u + \theta h, v + \theta k). \quad (6\text{–}1)$$

Then the distance D of a point $Q(\mathbf{x})$ on the surface near $P(\mathbf{x}_0)$ to the tangent plane is (Fig. 2–20):

$$D = (\mathbf{x} - \mathbf{x}_0) \cdot \mathbf{N} = \tfrac{1}{2}[h^2(\mathbf{x}_{uu} \cdot \mathbf{N}) + 2hk(\mathbf{x}_{uv} \cdot \mathbf{N}) + k^2(\mathbf{x}_{vv} \cdot \mathbf{N})]$$
$$+ \frac{1}{3!}\left(h\frac{\partial}{\partial u} + k\frac{\partial}{\partial v}\right)^3 \mathbf{N} \cdot \mathbf{x}(u + \theta h, v + \theta k),$$

of which the principal part is (compare with Eq. (5–9)):

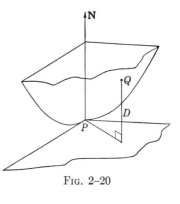

$$D_p = \tfrac{1}{2}(eh^2 + 2fhk + gk^2) = \tfrac{1}{2}\,\mathrm{II},$$
$$(h = du, \quad k = dv), \quad (6\text{–}2)$$

where D_p is positive or negative depending on whether Q lies on one or the other side of the tangent plane. Eq. (6–2) gives a geometrical illustration of II, comparable to the identification of ds^2 with I:

Fig. 2–20

When the second fundamental form does not vanish, it is equal to $2D_p$, where D_p is the principal part of the distance of the point $Q(u + du, v + dv)$ to the tangent plane at $P(u, v)$.

When $eg - f^2 > 0$, D_p (as well as D) retains its sign; when $eg - f^2 < 0$, it changes sign:

At an elliptic point the surface lies entirely on one side of the tangent plane, at a hyperbolic point it passes through the tangent plane in the asymptotic directions.

At a parabolic point there is one (asymptotic) direction, in which the contact is of higher order.

All types of points are illustrated on a torus (a surface obtained by rotating a circle about a line in its plane outside of the circle). The outside

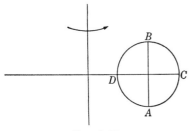

points (obtained by rotating BCA) are elliptic; the inside points (BDA) are hyperbolic; the circles obtained by rotating B and A have parabolic points (Fig. 2–21).

We can now ask for the directions in which the normal curvature is a maximum or minimum. We shall from now on write κ instead of κ_n, all through Chapter 2.

Fig. 2–21

The normal curvature in direction du/dv is given by Eqs. (5–6):

$$\kappa = \frac{e\,du^2 + 2f\,du\,dv + g\,dv^2}{E\,du^2 + 2F\,du\,dv + G\,dv^2} = \frac{e + 2f\lambda + g\lambda^2}{E + 2F\lambda + G\lambda^2} = \kappa(\lambda), \qquad (6\text{–}3)$$

where $\lambda = dv/du$. The extreme values of κ can be characterized by $d\kappa/d\lambda = 0$:

$$(E + 2F\lambda + G\lambda^2)(f + g\lambda) - (e + 2f\lambda + g\lambda^2)(F + G\lambda) = 0.$$

Since

$$E + 2F\lambda + G\lambda^2 = (E + F\lambda) + \lambda(F + G\lambda),$$
$$e + 2f\lambda + g\lambda^2 = (e + f\lambda) + \lambda(f + g\lambda),$$

we can, in this case, cast Eq. (6–3) into the simpler form:

$$\kappa = \frac{\text{II}}{\text{I}} = \frac{f + g\lambda}{F + G\lambda} = \frac{e + f\lambda}{E + F\lambda}. \qquad (6\text{–}4)$$

Hence κ satisfies the equations

$$(e - \kappa E)\,du + (f - \kappa F)\,dv = 0, \qquad (f - \kappa F)\,du + (g - \kappa G)\,dv = 0. \quad (6\text{–}4a)$$

Elimination of κ gives a quadratic equation for λ with real roots:

$$(Fg - Gf)\lambda^2 + (Eg - Ge)\lambda + (Ef - Fe) = 0 \qquad (6\text{–}5a)$$

or

$$\begin{vmatrix} dv^2 & -du\,dv & du^2 \\ E & F & G \\ e & f & g \end{vmatrix} = 0. \qquad (6\text{–}5b)$$

This equation determines two directions dv/du, in which κ obtains an extreme value, unless II vanishes or unless II and I are proportional. One value must be a maximum, the other a minimum. These directions are called the *directions of principal curvature*, or *curvature directions*. They are determined either by Eqs. (6–4a), (6–5a), or (6–5b). Since the roots λ_1, λ_2 satisfy the equation (compare with Eq. (6–5a)):

$$G\lambda_1\lambda_2 + F(\lambda_1 + \lambda_2) + E$$
$$= \frac{-1}{gF - Gf}[G(eF - Ef) - F(eG - Eg) - E(gF - Gf)] = 0,$$

the curvature directions are orthogonal (according to Eq. (2–10)). This also holds for the case that $gF - Gf = 0$, when, according to Eq. (6–5b), one of the directions is $du = 0$. We call the normal curvatures in the curvature directions the *principal curvatures*, and denote them by κ_1 and κ_2.

Integration of Eq. (6–5) gives us the *lines of curvature* on the surface, which form two sets of curves intersecting at right angles, or an *orthogonal family* of curves on the surface. From the existence theorem of ordinary differential equations, we can conclude that these curves cover the surface simply and without gaps in the neighborhood of every point where the coefficients of the first and second fundamental forms are continuous, except at the points where these coefficients are proportional. Such points are called *navel points* or *umbilics*, and we exclude them for the moment. Now let us take the lines of curvature as the parametric lines. Then Eqs. (6–5a, b) must be satisfied for $du = 0$, dv arbitrary, and for $dv = 0$, du arbitrary. Hence:

$$gF - Gf = 0, \qquad eF - Ef = 0.$$

In these equations $F = 0$ because the parametric lines are orthogonal; moreover, neither E nor G can be zero ($EG - F^2 > 0$). We therefore find that *when the parametric lines are lines of curvature:*

$$\boxed{F = 0, \qquad f = 0.} \tag{6–6}$$

This condition is necessary and, because of Eq. (6–5b), also sufficient. Now Eq. (6–3) takes the form:

$$\kappa = \frac{e\,du^2 + g\,dv^2}{E\,du^2 + G\,dv^2} = e\left(\frac{du}{ds}\right)^2 + g\left(\frac{dv}{ds}\right)^2. \tag{6–7}$$

This formula can be cast into a simple form. We find by substituting first $dv = 0$, then $du = 0$, into Eq. (6–7), that

$$\kappa_1 = \frac{e}{E}, \qquad \kappa_2 = \frac{g}{G}, \tag{6–8}$$

and if we introduce the angle α between the direction dv/du and the curvature direction $\delta v = 0$, we find from Eq. (2–9) that

$$\cos \alpha = \frac{E\,du\,\delta u}{ds\sqrt{E\,\delta u^2}} = \sqrt{E}\,\frac{du}{ds}, \qquad \sin \alpha = \sqrt{G}\,\frac{dv}{ds}. \tag{6–9}$$

Eq. (6–7) therefore takes the form:

$$\boxed{\kappa = \kappa_1 \cos^2 \alpha + \kappa_2 \sin^2 \alpha.} \tag{6–10}$$

This relation, which expresses the normal curvature in an arbitrary direction in terms of κ_1, κ_2, is known as *Euler's theorem.* Together with Meusnier's formula it gives full information concerning the curvature of any curve through P on the surface.

This theorem is one of the contributions to the theory of surfaces due to Leonard Euler (1707–1783), who enriched mathematics in an endless number of ways. It was published, with a proof different from ours, in the *Recherches sur la courbure des surfaces*, Mémoires de l'Academie des Sciences de Berlin 16 (1760), published 1767, pp. 119–143, one of the first papers on the theory of surfaces. Our proof follows that given by C. Dupin (see Section 2–7). Euler also made a start with the theory of developable surfaces in *De solidis, quorum superficiem in planum explicare licet*, Novi Comment. Acad. Petropol., 16, 1771, pp. 3–34. Here Euler introduced x, y, and z as functions of two parameters. This is the first, or one of the first, papers where curvilinear coordinates on a surface are used.

At an *umbilic* the coefficients of I and II are proportional, so that Eq. (6–8) gives $\kappa_1 = \kappa_2$, which means that all normal curvature vectors coincide (this certainly holds for real surfaces; see, however, Section 5–6). At an umbilic the directions of curvature are no longer given by Eq. (6–5), and we have to study the behavior of the surface with the aid of derivatives of higher order than the second. The number of umbilics on a surface is in general finite; on an ellipsoid, for instance, there are four (real) umbilics (see Exercise 12, Section 2–8), provided the axes are unequal. On a sphere all points are umbilics (see Section 3–5 for the converse theorem).

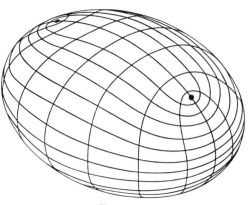

The behavior of the lines of curvature near an umbilic can be studied in Darboux, *Surfaces IV, Note VII*. There are several possibilities. An example is offered by the lines of curvature on an ellipsoid, where they curve around an umbilic like confocal conics (Fig. 2–22). See also A. Gullstrand, *Acta mathematica* **29**, 1905, p. 59.

A special case of umbilic is the umbilic at a

Fig. 2–22

parabolic point, or *parabolic umbilic*, where the coefficients e, f, and g all vanish. All points of the plane are parabolic umbilics (see Section 3–5 for the converse theorem). An example of an isolated umbilic of this kind can be constructed by taking a saddle point (Section 2–6), not with two upward and two downward slopes on the "height of land," but with three of them. In this case we can descend in directions at an angle of 120°, and ascend in the bisecting directions. Here every normal section has a point of inflection, so that $\kappa = 0$ in all directions. Such a point has been called a *monkey saddle*, since there are two downward slopes for the legs, leaving one slope for the tail (Fig. 2–23).

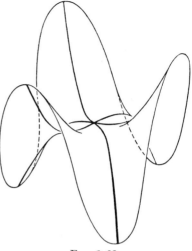

This monkey saddle (Affensattel) is described in D. Hilbert-S. Cohn Vossen, *Anschauliche Geometrie*, Berlin: Springer, 1932, which in Chapter IV, pp. 152–239, contains a highly original approach to the study of curves and surfaces, stressing the visual and imaginative side.

2–7 Dupin's indicatrix. The principal curvatures κ_1 and κ_2 can be found from Eq. (6–4) by substituting the values for λ found by

FIG. 2–23

solving Eq. (6–5a). A short cut is found by observing that these values of κ satisfy the two equations

$$(E\kappa - e) + (F\kappa - f)\lambda = 0, \qquad (F\kappa - f) + (G\kappa - g)\lambda = 0,$$

which can be simultaneously satisfied if and only if

$$\begin{vmatrix} E\kappa - e & F\kappa - f \\ F\kappa - f & G\kappa - g \end{vmatrix} = 0. \tag{7–1}$$

This quadratic equation in κ has κ_1 and κ_2 as roots. From this equation we derive

$$M = \frac{1}{2}(\kappa_1 + \kappa_2) = \frac{Eg - 2fF + eG}{2(EG - F^2)}, \quad \textit{the mean curvature}, \tag{7–2}$$

and

$$K = \kappa_1\kappa_2 = \frac{eg - f^2}{EG - F^2}, \quad \begin{matrix} \textit{the Gaussian curvature} \text{ (sometimes} \\ \text{called } \textit{total curvature)}. \end{matrix} \tag{7–3}$$

Eq. (7–3) shows that κ_1, κ_2 have the same sign at an elliptic point, and different signs at a hyperbolic point. The total curvature is zero at a parabolic point.

We can now give a simple diagram to illustrate Euler's theorem. Let us take an elliptic point and take both κ_1, $\kappa_2 > 0$. Consider the ellipse (Fig. 2–24)

$$\kappa_1 x^2 + \kappa_2 y^2 = 1, \qquad R_1 = \kappa_1^{-1},$$
$$R_2 = \kappa_2^{-1},$$

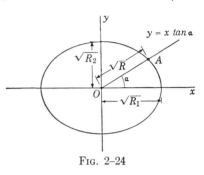

with principal semi-axes $\sqrt{R_1}$, $\sqrt{R_2}$. A line $y = x \tan \alpha$ through the center intersects the ellipse in the points

$$x = \frac{\pm \cos \alpha}{\sqrt{\kappa_1 \cos^2 \alpha + \kappa_2 \sin^2 \alpha}},$$
$$y = \frac{\pm \sin \alpha}{\sqrt{\kappa_1 \cos^2 \alpha + \kappa_2 \sin^2 \alpha}}.$$

Fig. 2–24

The distance OA intercepted by the ellipse on this line is

$$OA = \sqrt{x^2 + y^2} = \frac{1}{\sqrt{\kappa_1 \cos^2 \alpha + \kappa_2 \sin^2 \alpha}}.$$

Hence

$$\frac{1}{OA} = \sqrt{\kappa}, \quad \text{or} \quad OA = \sqrt{R}.$$

Here R is the radius of (normal) curvature in the direction which makes an angle α with the X-axis.

This ellipse is called the *indicatrix of Dupin*. If the X-axis represents one of the directions of curvature, then the distance of any point on the ellipse to the center is the square root of the radius of curvature in the corresponding direction on the surface.

When the point is hyperbolic, we can take $\kappa_1 > 0$, $\kappa_2 < 0$. The indicatrix of Dupin in this case is the set of conjugate hyperbolas (Fig. 2–25):

$$\kappa_1 x^2 + \kappa_2 y^2 = \pm 1, \qquad R_1 = \kappa_1^{-1},$$
$$R_2 = \kappa_2^{-1}.$$

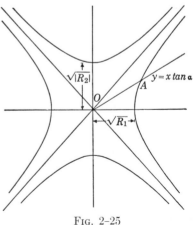

The length OA on a line making an angle α with the X-axis (corresponding to a direction of curvature) is now equal to $\sqrt{|R|}$. The two asymptotes of the hyperbolas represent the directions with $\kappa = 0$, or the *asymptotic directions* (hence

Fig. 2–25

the name). *They are real for hyperbolic points, imaginary for elliptic points.*

The indicatrix of a parabolic point, where

$$\kappa_2 = 0, \qquad \kappa = \kappa_1 \cos^2 \varphi, \qquad R_1 = \kappa_1^{-1}$$

is a pair of parallel lines in the direction of the single asymptotic line (Fig. 2–26).

We now intersect the surface with a plane, parallel to the tangent plane at an elliptic point, but only a small distance ϵ away from it. We project the intersection on the tangent plane. Then the principal part of the intersection, according to Eq. (6–2), will be given by the equation:

$$\frac{e}{E} x^2 + \frac{2f}{\sqrt{EG}} xy + \frac{g}{G} y^2 = 2\epsilon \qquad (x = \sqrt{E}\, h = \sqrt{E}\, du, \, y = \sqrt{G}\, k = \sqrt{G}\, dv).$$

If we measure x and y in the directions of curvature, this quadratic equation becomes (see Eq. (6–6)), after a similarity transformation $x \to x\sqrt{2\epsilon},\, y \to y\sqrt{2\epsilon}$:

$$\frac{e}{E} x^2 + \frac{g}{G} y^2 = 1, \qquad \kappa_1 x^2 + \kappa_2 y^2 = 1.$$

We thus have shown for an elliptic point:

The intersection of the surface with a plane close to the tangent plane and parallel to it is in a first approximation similar to the Dupin indicatrix.

In the case of a hyperbolic point we must intersect the surface with two planes close to the tangent plane, parallel to it, and one on each side. *The projection of the intersections of these planes with the surface on the tangent plane is again, in a first approximation, similar to the Dupin indicatrix.*

In the case of a parabolic point we need only one plane parallel to the tangent plane *to get, in a first approximation, the two parallel lines of the indicatrix.*

This theory has a meaning only at points where the second fundamental form does not vanish identically, that is, at points where the tangent plane has ordinary contact with the surface. At a monkey saddle the indicatrix, constructed according to the preceding rule, is given by Fig. 2–27.

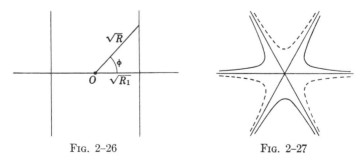

FIG. 2–26   FIG. 2–27

The asymptotes of the Dupin indicatrix correspond to the asymptotic directions on the surface, the axes of the indicatrix to the directions of curvature.   Hence:

The directions of curvature bisect the asymptotic directions.

Charles Dupin (1784–1873), a pupil of Monge at the Paris Ecole Polytechnique, wrote the *Développements de géometrie* (1813), in which he published a number of results which carry his name, as well as the treatment of lines of curvature and asymptotic lines which now is usually presented.   The subtitle of his book *Avec des applications à la stabilité des vaisseaux, aux déblais et remblais, au défilement, à l'optique, etc.* shows how the author never lost contact with engineering practice.   This is also evident in his second book, *Applications de géometrie* (1822).   Entering the Napoleonic navy as an engineer, Dupin lived to be a promoter of science and industry, a peer of France and a senator under Napoleon III.   The notion of mean curvature was introduced by Sophie Germain, Crelle's *Journal für Mathem.* 7, 1831, pp. 1–29.

2–8 Some surfaces.   1. *Sphere.*   Here (see Eq. (2–12)):

$$ds^2 = a^2 \, d\theta^2 + a^2 \cos^2 \theta \, d\varphi^2, \qquad u = \theta, \qquad v = \varphi,$$
$$\mathbf{x}_{uu}(-a \cos \theta \cos \varphi, \quad -a \cos \theta \sin \varphi, \quad -a \sin \theta),$$
$$\mathbf{x}_{uv}(\quad a \sin \theta \sin \varphi, \quad -a \sin \theta \cos \varphi, \qquad 0 \quad),$$
$$\mathbf{x}_{vv}(-a \cos \theta \cos \varphi, \quad -a \cos \theta \sin \varphi, \qquad 0 \quad),$$

$$\sqrt{EG - F^2} = a^2 \cos \theta,$$

$$e = \frac{\begin{vmatrix} -a \cos \theta \cos \varphi & -a \cos \theta \sin \varphi & -a \sin \theta \\ -a \sin \theta \cos \varphi & -a \sin \theta \sin \varphi & a \cos \theta \\ -a \cos \theta \sin \varphi & +a \cos \theta \cos \varphi & 0 \end{vmatrix}}{a^2 \cos \theta} = \frac{a^3 \cos \theta}{a^2 \cos \theta} = a.$$

Similarly, $$f = 0, \qquad g = a \cos^2 \theta,$$

$$\mathrm{II} = a(d\theta^2 + \cos^2 \theta \, d\varphi^2), \tag{8–1}$$

$$\kappa = \frac{\mathrm{II}}{\mathrm{I}} = + \frac{1}{a}.$$

This is the case mentioned at the end of Section 2–6, where I and II are proportional.   *All points on the sphere are umbilics,* and since the principal directions are undetermined at all points *all curves on the sphere may be taken as lines of curvature.*   The + sign of κ is due to the choice of

$$\mathbf{N}(-\cos \theta \cos \varphi, \; -\cos \theta \sin \varphi, \; -\sin \theta),$$

which means that $\mathbf{N}$ is directed toward the center of the sphere.   This is also the sense of $\kappa\mathbf{N}$, the curvature vector of a normal section.

The asymptotic lines of the sphere satisfy the equation

$$d\theta^2 + \cos^2\theta \, d\varphi^2 = 0, \tag{8-2}$$

and are therefore imaginary curves; since I and II are proportional they are also isotropic curves. Conversely, all isotropic curves on the sphere are at the same time asymptotic curves. Waiving for a moment our restriction to real figures only, we find the integral of Eq. (8-2) in the form

$$\tan\left(\frac{\theta}{2} + \frac{\pi}{4}\right) = ce^{\pm\, i\varphi}, \tag{8-3}$$

which in the case of the $+$ sign leads to

$$\sin\theta = \frac{c^2 e^{2i\varphi} - 1}{c^2 e^{2i\varphi} + 1}, \qquad \cos\theta = \frac{2ce^{i\varphi}}{c^2 e^{2i\varphi} + 1}, \qquad u = e^{2i\varphi}.$$

Inserting these values into the Eqs. (1-7) for x, y, z, we obtain

$$x + iy = a\,\frac{2cu}{c^2 u + 1}, \qquad z = a\,\frac{c^2 u - 1}{c^2 u + 1}.$$

Elimination of u shows that these curves are plane, and since they are isotropic, they must be straight lines (Section 1-12).

We obtain a similar result when in Eq. (8-3) we take the $-$ sign:

Through every point of the sphere pass two isotropic straight lines lying on the sphere. They are the asymptotic lines of the sphere.

The Dupin indicatrix at a point of the sphere is a circle.

2. *Surface of revolution.* Here (see Eq. (2-14)):

$$ds^2 = (1 + f'^2)\, dr^2 + r^2\, d\varphi^2.$$

We find

$$e = \frac{f''}{\sqrt{1 + f'^2}}, \qquad f = 0,{}^* \qquad g = \frac{rf'}{\sqrt{1 + f'^2}}.$$

We conclude that because of $f = F = 0$ the coordinate lines are lines of curvature (see Eq. (6-6)).

The lines of curvature of a surface of revolution are its meridians and parallels.

The asymptotic lines are the integral curves of the equation

$$f''\, dr^2 + rf'\, d\varphi^2 = 0$$

---

* The reader will have little trouble to distinguish between the f of the second fundamental form and the f of $f(r)$.

and can therefore be obtained by a quadrature. Since

$$eg - f^2 = \frac{rf'f''}{1 + f'^2},$$

the points of a real surface are elliptic or hyperbolic, depending on whether $f'f'' > 0$ or < 0 respectively. The circles of parabolic points are given by $f' = 0$, which gives the points which in the profile have tangents $\perp Z$-axis and $f'' = 0$, which gives the points of inflection of the profile. Between these circles the asymptotic lines are alternately real or imaginary (when $f' = 0$ and $f'' = 0$ are not satisfied at the same point of the profile).

When $f'' = 0$ at all points of the surface we have $f = ar + b$: a *circular cone* (Eqs. (1–8, 2–16)). All points are parabolic except the vertex $r = 0$, where $e = f = g = 0$ (this point is singular, not umbilic).

As soon as f'' *contains a factor* r, the point $r = 0$ has also $e = f = g = 0$. This happens, for example, when $f = ar^4$. The profile at the point $r = 0$ has a tangent line with a contact of order 3 in r.

The asymptotic lines are *orthogonal* when (see Exercise 6, Section 2–3):

$$E(rf') + G(f'')$$
$$= rf'(1 + f'^2) + r^2f'' = 0.$$

When we integrate this equation, we obtain:

$$f' = \frac{a}{\sqrt{r^2 - a^2}}, \quad \text{or} \quad f = a \cosh^{-1}\frac{r}{a} + c,$$
$$(a, c \text{ integration constants}),$$

which means that the profile curve is a catenary. This surface, obtained by rotating the curve

$$y = a \cosh^{-1}\frac{x}{a}$$

or

$$x = a \cosh\frac{y}{a} = \frac{a}{2}\left(e^{\frac{y}{a}} + e^{-\frac{y}{a}}\right)$$

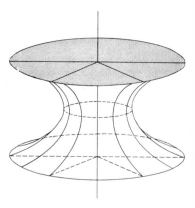

Fig. 2–28

about the Y-axis, is called a *catenoid* (Fig. 2–28). Surfaces on which the asymptotic lines form an orthogonal net are called *minimal surfaces*. We have thus found that *the only surfaces of revolution which are also minimal surfaces are catenoids*.

3. *Right conoid.* Here (Section 2–2):

$$ds^2 = dr^2 + (r^2 + f'^2)\, d\varphi^2, \qquad f = f(\varphi).$$

We find

$$e = 0, \qquad f = \frac{-f'}{\sqrt{r^2 + f'^2}}, \qquad g = \frac{f''r}{\sqrt{r^2 + f'^2}}.$$

The asymptotic lines are given by

$$-2f'\, dr\, d\varphi + f''r\, d\varphi^2 = 0.$$

Since $eg - f^2 < 0$, all points are hyperbolic and the asymptotic lines are real. One set is given by $d\varphi = 0$, which means that the straight lines are asymptotic lines. The other set is the solution of

$$-2f'\, dr + rf''\, d\varphi = 0, \qquad \frac{dr}{r} = \frac{f''\, d\varphi}{2f'}, \qquad cr^2 = f'$$

and can therefore always be found in finite form. The two sets are orthogonal when $f'' = 0$ or $f = a\varphi + b$, which characterizes the right helicoid (Eq. (2–19)).

The right helicoid is the only right conoid which is also a minimal surface. In this case the curved asymptotic lines are given by $r = $ constant; they are circular helices (Fig. 2–29). Bisecting the angle of these curves (Section 2–3, Exercise 9) are the lines of curvature:

$$-a\, dr^2 + a(r^2 + a^2)\, d\varphi^2 = 0; \qquad \varphi + c_1 = \pm a \sinh^{-1} \frac{r}{a}.$$

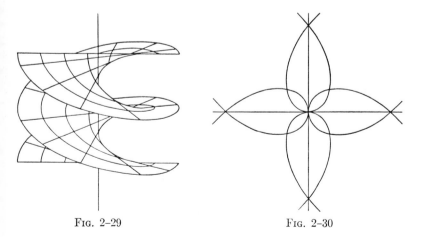

FIG. 2–29   FIG. 2–30

Since r and φ are the polar coordinates of the projection of a point of the conoid on the XOY-plane, the lines of curvature are projected on the XOY-plane in the curves (Fig. 2–30):

$$r = \pm a \sinh \frac{1}{a} (\varphi + c_1).$$

4. *Developable surfaces.* For a tangent developable:

$$\mathbf{y}(s, v) = \mathbf{x}(s) + v\mathbf{t}(s). \tag{8–4}$$

Here

$$\mathbf{y}_s = \mathbf{t} + v\kappa\mathbf{n}, \qquad \mathbf{y}_v = \mathbf{t},$$
$$\mathbf{y}_{ss} = \kappa\mathbf{n} + v(\kappa\mathbf{n})', \qquad \mathbf{y}_{sv} = \kappa\mathbf{n}, \qquad \mathbf{y}_{vv} = 0,$$

$$\mathbf{N} = \frac{-v\kappa\mathbf{b}}{\sqrt{EG - F^2}} = -\mathbf{b},$$

$$E = 1 + v^2\kappa^2, \qquad F = 1, \qquad G = 1,$$
$$e = -v\kappa\tau \qquad f = 0, \qquad g = 0.$$

We conclude that $eg - f^2 = 0$, so that there are only parabolic points. The asymptotic lines are the solution of $eds^2 = 0$, hence they are the straight lines of the surface. The lines of curvature are the integral curves of the equation

$$\begin{vmatrix} dv^2 & -ds\,dv & ds^2 \\ E & 1 & 1 \\ e & 0 & 0 \end{vmatrix} = 0,$$

or

$$ds^2 + ds\,dv = 0,$$

hence they are the straight lines and the curves $s + v = c$, which are the involutes of the edge of regression (Section 1–11).

When, in Eq. (8–4), $\mathbf{x}(s)$ reduces to a constant vector $\mathbf{a}$, then Eq. (8–4) represents a cone of which the vertex is the end point of $\mathbf{a}$:

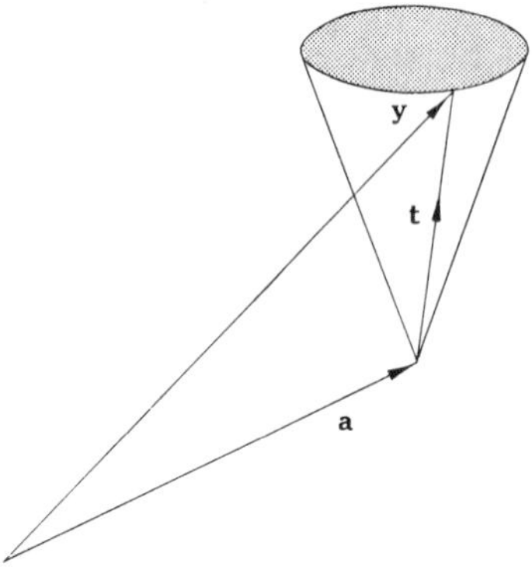

Fig. 2–31

$$\mathbf{y} = \mathbf{a} + v\mathbf{t}(s),$$

where $\mathbf{t}$ is now a given unit vector field, determined for instance by a curve on a unit sphere (Fig. 2–31). Here again we find that $eg - f^2 = 0$, and that the lines of curvature are formed by the generating lines and their orthogonal trajectories.

A cylinder is given by the equation

$$\mathbf{y} = \mathbf{x}(s) + v\mathbf{a}$$

where $\mathbf{a}$ is a constant vector. Here we also find that $eg - f^2 = 0$.

We shall now prove that $eg - f^2 = 0$ is not only a necessary, but also a sufficient condition that a surface be developable.

We prove this by means of the identity

$$eg - f^2 = (\mathbf{x}_u \cdot \mathbf{N}_u)(\mathbf{x}_v \cdot \mathbf{N}_v) - (\mathbf{x}_u \cdot \mathbf{N}_v)(\mathbf{x}_v \cdot \mathbf{N}_u) = (\mathbf{x}_u \times \mathbf{x}_v) \cdot (\mathbf{N}_u \times \mathbf{N}_v)$$
$$= (\mathbf{N}\mathbf{N}_u\mathbf{N}_v)\sqrt{EG - F^2}, \tag{8–5}$$

which shows that $eg - f^2 = 0$ is identical with $(\mathbf{N}\mathbf{N}_u\mathbf{N}_v) = 0$. This can happen either (a) when $\mathbf{N}_u$ or $\mathbf{N}_v$ vanishes, or (b) when $\mathbf{N}_u$ is collinear with $\mathbf{N}_v$ ($\mathbf{N}$ is perpendicular to $\mathbf{N}_u$ and $\mathbf{N}_v$). In case (a) $\mathbf{N}$ depends on only one parameter and the surface is the envelope of a family of ∞^1 planes, and hence (see Section 2–4) a developable. In case (b) we take as one set of coordinate curves on the surface the asymptotic curves with equation

$$e\,du^2 + 2f\,du\,dv + g\,dv^2 = (\sqrt{e}\,du + \sqrt{g}\,dv)^2 = 0.$$

If these curves are taken as the curves $v = $ constant in the new coordinate system, then $e = f = 0$ or $\mathbf{x}_u \cdot \mathbf{N}_u = \mathbf{x}_v \cdot \mathbf{N}_u = 0$, hence $\mathbf{N}_u = 0$, which brings us back to case (a). The theorem can also be stated as follows (see Eq. (7–3)):

A necessary and sufficient condition that a surface be developable is that the Gaussian curvature vanish.

EXERCISES

1. What is the second fundamental form when the surface is given by the equation $z = f(x, y)$?

2. The differential equation of the developable surfaces $z = f(x, y)$ is $rt - s^2 = 0$. (Here and in the following text we often use the notation

$$p = \partial z/\partial x, \qquad q = \partial z/\partial y, \qquad r = \partial^2 z/\partial x^2, \qquad s = \partial^2 z/\partial x \partial y, \qquad t = \partial^2 z/\partial y^2.)$$

3. The differential equation of all minimal surfaces $z = f(x, y)$ is

$$t(1 + p^2) - 2pqs + r(1 + q^2) = 0.$$

4. What is the geometrical meaning of the general integral of the partial differential equations

$$\text{(a) } xp + yq = 0, \qquad \text{(b) } yp - xq = 0?$$

5. Find the equation of the lines of curvature in the case that the equation of the surface is $z = f(x, y)$.

6. Show that the surface

$$z = a + bx + cy + \sum_{n=2}^{N} a_n(px + qy)^n,$$

where the coefficients of x and y are all constants, is developable.

7. The sum of the normal curvatures at a point of a surface in any pair of orthogonal directions is constant.

8. Find the radii of principal curvature at a point of a surface of revolution $x = u \cos v$, $y = u \sin v$, $z = f(u)$, and interpret the answer geometrically.

9. Find the radii of principal curvature at a point of a developable surface.

10. The asymptotic lines on the surface

$$z = \frac{x^4}{a^4} - \frac{y^4}{b^4}$$

are the curves in which the surface is met by the two families of cylinders

$$\frac{x^2}{a^2} + \frac{y^2}{b^2} = \text{constant}, \qquad \frac{x^2}{a^2} - \frac{y^2}{b^2} = \text{constant}.$$

11. *Asymptotic lines on surfaces of revolution.* Show that the surface of revolution with the profile $z = f(r)$, given in Table A, has asymptotic lines of which the projection on the XOY-plane has the equation $r = r(\varphi)$, given in Table B, where r, φ are polar coordinates.

Table A	Table B
(a) $z = \sqrt{r}$	(a) $\varphi + c = \frac{1}{2}\sqrt{2}\ln r$
(b) $z = 6r^{-1}$	(b) $\varphi + c = \sqrt{2}\ln r$
(c) $z = 6 \ln r$	(c) $\varphi + c = \ln r$
(d) $z = (r - a)^2$	(d) $r = a\cos^2(\varphi/2 + c)$
(e) $z = \left[r\sqrt{a^2 - r^2} - a^2 \cos^{-1}\dfrac{r}{a} \right]$	(e) $\varphi + c = \cos^{-1}\dfrac{r}{a}.$

Plaster models of these surfaces and of some others (twelve in all) were constructed in 1885 at the University of Munich under the supervision of A. Brill and can be inspected in many cabinets of mathematical models.

12. Find the umbilics of the ellipsoid and prove that the tangent planes at these points are parallel to the circular sections of the ellipsoid (that is, to those planes which intersect the ellipsoid in circles).

13. Find the parabolic curves on a torus (Fig. 2–21) by computation.

14. Show that the asymptotic directions of a surface, given by $F(x, y, z) = 0$, are given by

$$dx \, dF_x + dy \, dF_y + dz \, dF_z = 0, \qquad F_x \, dx + F_y \, dy + F_z \, dz = 0.$$

15. Show that the directions of curvature in the case of Exercise 14 are given by

$$\begin{vmatrix} F_x & dF_x & dx \\ F_y & dF_y & dy \\ F_z & dF_z & dz \end{vmatrix} = 0, \qquad F_x \, dx + F_y \, dy + F_z \, dz = 0.$$

16. Show that a surface $F(x, y, z) = 0$ is developable if

$$\begin{vmatrix} F_{xx} & F_{xy} & F_{xz} & F_x \\ F_{yx} & F_{yy} & F_{yz} & F_y \\ F_{zx} & F_{zy} & F_{zz} & F_z \\ F_x & F_y & F_z & 0 \end{vmatrix} = 0, \quad F_x = \partial F/\partial x, \text{ etc.}$$

17. *Cylindroid.* The locus of the end points of the curvature vectors at a point P of a surface belonging to all curves passing through P is a right conoid with the equation

$$z(x^2 + y^2) = \kappa_1 x^2 + \kappa_2 y^2.$$

This surface is called *conoid of Plücker* or cylindroid; the surface normal is its double line (Fig. 2–32).

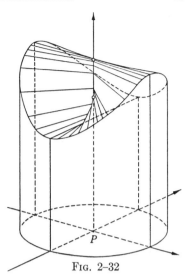

Fig. 2–32

2–9 A geometrical interpretation of asymptotic and curvature lines. The lines of curvature can be characterized by the following property, due to Monge.

A necessary and sufficient condition that a curve on a surface be a line of curvature is that the surface normals along this curve form a developable surface.

To prove this theorem, let the curve be given by $\mathbf{x} = \mathbf{x}(s)$. Then

$$(d\mathbf{x}/ds) \cdot \mathbf{N} = \mathbf{t} \cdot \mathbf{N} = 0.$$

A point on the ruled surface formed by the surface normals can be given by

$$\mathbf{y} = \mathbf{x}(s) + u\mathbf{N}(s), \tag{9–1}$$

where u is the distance of point $\mathbf{y}$ to point $\mathbf{x}$. Then

$$\mathbf{y}_s = \mathbf{t} + u\mathbf{N}_s, \quad \mathbf{y}_u = \mathbf{N}, \quad \mathbf{y}_{su} = \mathbf{N}_s, \quad \mathbf{y}_{uu} = 0,$$

hence $g = 0$, and the condition of developability $eg - f^2 = 0$ requires that $f = 0$ or

$$(\mathbf{t} \, \mathbf{N} \, \mathbf{N}_s) = 0.$$

$\mathbf{N}$ is perpendicular to $\mathbf{t}$ and $\mathbf{N}_s$, so that the only ways in which this condition can be satisfied is (a) $\mathbf{N}_s = 0$, (b) $\mathbf{t}$ is collinear with $\mathbf{N}_s$. In the case (a) the surface normals form a cylinder or a plane, which are developable surfaces. Case (b) can be expressed as follows:

$$d\mathbf{N} = \lambda\, d\mathbf{x} \text{ (along the curve)}; \qquad \text{(a) } \lambda = 0, \quad \text{(b) } \lambda \neq 0. \qquad (9\text{-}2)$$

Eq. (9-2) is equivalent to

$$\mathbf{N}_u\, du + \mathbf{N}_v\, dv = \lambda(\mathbf{x}_u\, du + \mathbf{x}_v\, dv),$$

which vector equation is equivalent to the two following scalar equations, obtained by scalar multiplication with $\mathbf{x}_u$ and $\mathbf{x}_v$ respectively:

$$(e + \lambda E)\, du + (f + \lambda F)\, dv = 0,$$
$$(f + \lambda F)\, du + (g + \lambda G)\, dv = 0,$$

which in turn are equivalent to Eq. (6-4a), with $\lambda = -\kappa$; conversely, Eq. (6-4a) is equivalent to Eq. (9-2). The curves are therefore lines of curvature and only these; moreover, we have found a new equation to characterize these lines:

$$\boxed{d\mathbf{N} + \kappa\, d\mathbf{x} = 0,} \qquad (9\text{-}3)$$

where κ is the normal curvature in the direction $d\mathbf{x}$ of the line of curvature. This is the formula of *Rodrigues*.

Olinde Rodrigues (1794–1851), another of Monge's pupils, published his results in the *Correspondance sur l'Ecole Polytechnique* 3 (1815) and in the *Bull. Soc. Philomatique* 2 (1815). His name is also attached to a theorem in Legendre functions. He became a follower of St. Simon, and was an editor of the collected works of this reformer.

Eq. (9-2) with $\kappa = 0$ represents a cylinder or a plane, and with κ constant a cone. This can be verified by differentiating Eq. (9-1):

$$d\mathbf{y} = d\mathbf{x} + u(-\kappa\, d\mathbf{x}) + \mathbf{N}\, du = d\mathbf{x}(1 - \kappa u) + \mathbf{N}\, du,$$

which is zero when $u = \kappa^{-1}$, $du = 0$, and only in this case. The surface normals along a line of curvature, in all other cases, form a tangential developable, so that they are tangent to a space curve, the edge of regression. Each surface normal is tangent to two edges of regression; the points of tangency are at distances κ_1^{-1} and κ_2^{-1} from P; they are the *centers of principal curvature*. The two families of lines of curvature on the surface, forming an *orthogonal net*, thus determine two sets of developable surfaces intersecting at right angles along the surface normals. The *center surfaces* are the loci of the centers of principal curvature of the surface. A simple

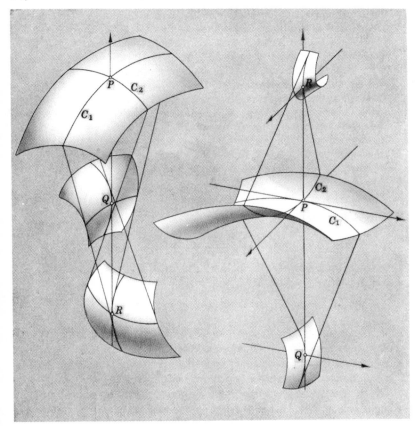

Fig. 2–33

example is offered by the meridians and parallels of a surface of revolution, which are lines of curvature because along the meridians the surface normals form planes, and along the parallels form cones.　The cones and planes are developable surfaces.　They are orthogonal to each other; one of the center surfaces here degenerates into the axis.　The general case is illustrated by Fig. 2–33, where C_1 and C_2 are the lines of curvature through P, and Q and R are the points on the center surfaces.　The edges of regression of the developable surfaces lie on the center surfaces.

Monge was led to the lines of curvature of a surface in his *Mémoire sur la théorie des déblais et des remblais*, Mémoires de l'Acad. des Sciences, 1781, pp. 666–704, which was based on the following engineering problem: To decompose two given equivalent volumes into infinitely small particles which correspond to each other in such a way that the sum of the products of the paths described by the transportation of each particle of the first volume (the "déblai") to its corresponding particle of the second volume (the "remblai") and the volume of the particle is a minimum. The orbits were supposed to be straight lines. This led to rectilinear congruences; that is, to families of straight lines with the property that through one point of a region of space one and only one line passes. These lines can always be arranged in two sets of developable surfaces; when these surfaces are normal, the lines of the congruence are normal to a set of surfaces on which the developable surfaces cut out the lines of curvature. See the monograph of P. Appell, *Le problème géométrique des déblais et des remblais*, Mémorial des sciences mathématiques, **29**, 1928, 34 pp.

The asymptotic lines are characterized by the equation II = 0 or, according to Eq. (5–8):

$$dx \cdot dN = 0. \qquad (9–4)$$

This equation can be written in the form $t \cdot dN = 0$, and since $t \cdot N = 0$, also in the form $N \cdot dt = 0$, or

$$\kappa N \cdot n = 0,$$

which is satisfied either by $\kappa = 0$ or by $N \cdot n = 0$. This leads to the following characterization of asymptotic lines:

All straight lines on a surface are asymptotic lines. Along a curved asymptotic line the osculating plane coincides with the tangent plane.

The converse is also true, and when we consider that any plane through a straight line may be considered as an osculating plane of the line, we can summarize our results as follows:

A necessary and sufficient condition that a curve on a surface be asymptotic is that the tangent plane of the surface coincide with the osculating plane of the curve.

As an example we can take the straight lines of a ruled surface, which here form one family of asymptotic lines (see our example of the developable surfaces and of the right conoids, Section 2–8). The hyperboloid of one sheet has two sets of straight lines, which are its asymptotic lines. The isotropic lines on the sphere are its asymptotic lines (Section 2–8).

2–10 Conjugate directions. In his study of the indicatrix which carries his name, Dupin pointed out that the directions on the surface corresponding to conjugate diameters in the indicatrix lead to curves with interesting properties. He called these directions *conjugate;* they determine *conjugate families of curves.*

The diameters $y = m_1x$ and $y = m_2x$ in the ellipse (Fig. 2–34)

$$\frac{x^2}{a^2} + \frac{y^2}{b^2} = 1$$

are called *conjugate* if one diameter d_1 is parallel to the tangents at the end points B_1, B_2 of the other diameter.   Since the tangent at $B_1(x_1, m_2x_1)$ has the equation

$$\frac{xx_1}{a^2} + \frac{m_2yx_1}{b^2} = 1,$$

we have

$$m_1 = -b^2/a^2m_2,$$

or

$$m_1m_2 = -b^2/a^2.$$

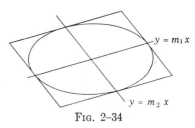

The relation between two conjugate diameters is reciprocal: the tangents at the end points A_1, A_2 of d_1 are parallel to d_2.   For the hyperbolas

FIG. 2–34

$$\frac{x^2}{a^2} - \frac{y^2}{b^2} = \pm 1$$

the condition is $m_1m_2 = b^2/a^2$.

––––––––––

Conjugate directions with respect to the indicatrix, both in the elliptic and in the hyperbolic case, are related by the equation

$$m_1m_2 = -\frac{\kappa_1}{\kappa_2}.$$

Since $m = \tan \varphi = \sqrt{\frac{G}{E}} \frac{dv}{du}$ (see Eq. (6–9)) and $\frac{\kappa_1}{\kappa_2} = \frac{e}{E} \frac{G}{g}$ (see Eq. (6–8)), we find for the equation determining conjugate directions:

$$\sqrt{\frac{G}{E}} \frac{dv}{du} \sqrt{\frac{G}{E}} \frac{\delta v}{\delta u} = -\frac{G}{E} \frac{e}{g},$$

or

$$e\, du\, \delta u + g\, dv\, \delta v = 0. \tag{10–1}$$

This equation for conjugate directions is obtained for the special case that $f = 0$.   To obtain expressions valid for the general case, in which $f \neq 0$, we substitute for e, f, g, their values (5–7a).   Eq. (10–1) then can be written (adding $f\, \delta u\, dv$ and $f\, du\, \delta v$, which are here zero):

$$\mathbf{x}_u \cdot \mathbf{N}_u\, du\, \delta u + \mathbf{x}_v \cdot \mathbf{N}_v\, dv\, \delta v + \mathbf{x}_u \cdot \mathbf{N}_v\, du\, \delta v + \mathbf{x}_v \cdot \mathbf{N}_u\, dv\, \delta u = 0,$$

or

$$(\mathbf{x}_u\, du + \mathbf{x}_v\, dv) \cdot (\mathbf{N}_u\, \delta u + \mathbf{N}_v\, \delta v) = 0,$$

or

$$dx \cdot \delta \mathbf{N} = 0. \qquad (10\text{--}2a)$$

This equation may also be written

$$\delta \mathbf{x} \cdot d\mathbf{N} = 0. \qquad (10\text{--}2b)$$

In an arbitrary curvilinear coordinate system this equation takes the form

$$e \, du \, \delta u + f(du \, \delta v + dv \, \delta u) + g \, dv \, \delta v = 0. \qquad (10\text{--}3)$$

Since asymptotic lines are defined by $dx \cdot d\mathbf{N} = 0$ we see that *asymptotic lines are self-conjugate.* This agrees with the fact that the asymptotes of the indicatrix are self-conjugate diameters. One family of curves may be selected arbitrarily; it determines the conjugate family.

Eqs. (10–2a, b) allow a simple geometrical interpretation. The tangent planes along a curve $C(\mathbf{x}(s))$ on the surface form a developable surface. Let the equation

$$\mathbf{X} = \mathbf{x} + u\mathbf{p}$$

indicate the generators of this surface. Along the generator the unit surface normal vector is $\mathbf{N}$, so that from $\mathbf{N} \cdot d\mathbf{X} = 0$, where

$$d\mathbf{X} = dx + u \, d\mathbf{p} + \mathbf{p} \, du$$

it follows that $0 = \mathbf{N} \cdot dx + u\mathbf{N} \cdot d\mathbf{p} + \mathbf{N} \cdot \mathbf{p} \, du$. Since $\mathbf{N} \cdot dx = 0$ and $\mathbf{N} \cdot \mathbf{p} = 0$, we conclude that

$$\mathbf{N} \cdot d\mathbf{p} = 0,$$

which leads to

$$d\mathbf{N} \cdot \mathbf{p} = 0.$$

If we denote the direction of $\mathbf{p}$ by $\delta \mathbf{x}$, we find that

$$\delta \mathbf{x} \cdot d\mathbf{N} = 0,$$

which means that the directions of dx and $\delta \mathbf{x}$ are conjugate. In other words:

The generating lines of the developable surface enveloped by the tangent planes to a surface along a curve C on the surface have along C a direction conjugate to the direction of C. In simpler, but less exact, words:

The line of intersection of two tangent planes at consecutive points in a direction on a surface has the conjugate direction. This relation is reciprocal.

The lines of curvature form a conjugate set. This follows from the properties of the indicatrix (its axes are conjugate). It also follows from the orthogonality relation $d\mathbf{x} \cdot \delta\mathbf{x} = 0$, if we take $d\mathbf{x}$ and $\delta\mathbf{x}$ to be in the directions of curvature. In this case, we obtain from Rodrigues' formula $\delta\mathbf{x} = -\kappa_1^{-1}\,\delta\mathbf{N}$ that $d\mathbf{x} \cdot \delta\mathbf{N} = 0$.

When the parametric lines are conjugate, Eq. (10–3) is satisfied for $du = 0$, $\delta v = 0$, which means that

$$f = 0.$$

Conversely, when $f = 0$, the parametric lines are conjugate.

Since the lines of curvature are both orthogonal and conjugate, we have returned to Eq. (6–6), expressing that *the necessary and sufficient condition that the parametric lines are lines of curvature is:*

$$F = 0, \qquad f = 0.$$

EXAMPLE. Let us find the curves conjugate to the parallels $\theta = u =$ constant on the sphere (Section 2–2). They satisfy the equation

$$g\,dv\,\delta v = g\,d\varphi\,\delta\varphi = 0, \qquad \text{or} \qquad \delta\varphi = 0,$$

which means that the meridians are conjugate to the parallels. Indeed, the planes tangent along the parallels envelop cones whose generating lines are tangent to the meridians; the planes tangent along the meridians envelop cylinders whose generating lines are tangent to the parallels.

It is convenient to combine in a table the conditions satisfied by $E, F, G,$ e, f, g when the parametric curves belong to one or another of the specific types of curves which we have discussed. These curves are

orthogonal, when $F = 0$,
lines of curvature, when $F = 0, f = 0$,
conjugate, when $f = 0$,
isotropic, when $E = 0, G = 0$ (since in this case $\mathrm{I} = F\,du\,dv$),
asymptotic, when $e = 0, g = 0$ (since in this case $\mathrm{II} = f\,du\,dv$).

2–11 Triply orthogonal systems of surfaces. A point on a surface or in the plane is determined by two parameters, which form a curvilinear system of coordinates; sometimes this system may be rectilinear as in the case of cartesian coordinates in the plane. We can in a similar way determine a point in space by means of three parameters, or curvilinear coordinates. For this purpose we transform the cartesian coordinates (x, y, z) of a point P by means of the equations

$$x = x(u, v, w), \; y = y(u, v, w), \; z = z(u, v, w), \; \text{or } \mathbf{x} = \mathbf{x}(u, v, w) \text{ for short.}$$
$$(11\text{–}1)$$

The surfaces $u =$ constant, $v =$ constant, $w =$ constant are the *coordinate surfaces;* they intersect in the *coordinate lines.* Examples are the transformations to cylindrical or spherical coordinates, which can be written as follows:

(a) $\qquad\qquad x = u \cos v, \qquad y = u \sin v, \qquad z = w,$

(b) $\qquad x = u \cos v \cos w, \quad y = u \cos v \sin w, \quad z = u \sin v.$ $\qquad$ (11–2)

Since

$$dx = x_u \, du + x_v \, dv + x_w \, dw,$$

the square of its length takes the form

$$ds^2 = dx \cdot dx = x_u \cdot x_u \, du^2 + 2x_u \cdot x_v \, du \, dv + 2x_u \cdot x_w \, du \, dw + x_v \cdot x_v \, dv^2$$
$$+ 2x_v \cdot x_w \, dv \, dw + x_w \cdot x_w \, dw^2. \quad (11–3)$$

Of particular interest are those curvilinear coordinates for which

$$x_u \cdot x_v = x_v \cdot x_w = x_w \cdot x_u = 0. \qquad (11–4)$$

The coordinate lines, and therefore also the coordinate surfaces, are perpendicular. This orthogonality can be proved in the same way as in the two-dimensional case (Eq. (2–10)). Examples of orthogonal systems of curvilinear coordinates are (a) the cartesian, (b) the cylindrical, and (c) the spherical system of coordinates, where the respective coordinate surfaces are (a) planes, (b) planes and cylinders, and (c) planes, spheres, and cones. The coordinate surfaces, in this case, form a *triply orthogonal system of surfaces.* Every transformation (11–1), subject to the conditions (11–4) determines such a system. There exists no simple way of finding such systems, since it requires the solution of three partial differential equations (11–4) in three dependent and three independent variables. We shall therefore content ourselves with a few examples, of which the simplest have already been presented. Another example is the system expressed by the equation

$$\frac{x^2}{a^2 - \lambda} + \frac{y^2}{b^2 - \lambda} + \frac{z^2}{c^2 - \lambda} = 1, \quad a^2 < b^2 < c^2, \qquad (11–5)$$

which is a system of quadric surfaces with the origin as center. When $\lambda < a^2$ it represents ellipsoids, when $a^2 < \lambda < b^2$ it represents hyperboloids of one sheet, and when $b^2 < \lambda < c^2$, hyperboloids of two sheets. When $\lambda > c^2$ the surfaces are imaginary. For given (x, y, z) Eq. (11–5) represents a cubic equation in λ, which, as can be shown (see Exercise 7, Section 2–11), has three real roots, one $< a^2$, one between a^2 and b^2, and another between b^2 and c^2. *Through every point of space passes one ellipsoid, one hyperboloid of one sheet, and one hyperboloid of two sheets.* When $z = 0$ they cut out of the XOY-plane a set of conics with the same foci, since $b^2 - \lambda - (a^2 - \lambda) = b^2 - a^2$ is independent of λ. Such a system of conics is called a *confocal system.* For this reason we call Eq. (11–5) a *system of confocal quadrics*, although there are no points which can be called common foci.

The tangent plane at a point $P(x_1, y_1, z_1)$ to a quadric with the origin as its center has the equation

$$\frac{xx_1}{a^2 - \lambda} + \frac{yy_1}{b^2 - \lambda} + \frac{zz_1}{c^2 - \lambda} = 1.$$

When two quadrics passing through P are given by $\lambda_1, \lambda_2, \lambda_1 \neq \lambda_2$, then they form an angle α of which the cosine is given by a fraction with the numerator:

$$\frac{x_1^2}{(a^2 - \lambda_1)(a^2 - \lambda_2)} + \frac{y_1^2}{(b^2 - \lambda_1)(b^2 - \lambda_2)} + \frac{z_1^2}{(c^2 - \lambda_1)(c^2 - \lambda_2)}.$$

But $x_1 y_1 z_1$ satisfy also Eq. (11–5) for $\lambda = \lambda_1$ and $\lambda = \lambda_2$, hence we find that

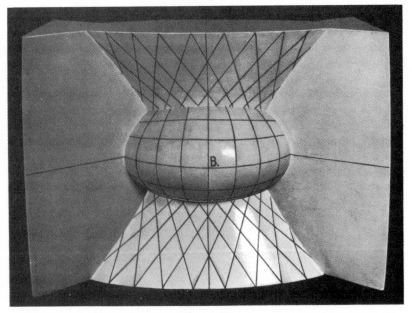

Fig. 2–35

$$x_1^2\left[\frac{1}{a^2-\lambda_1}-\frac{1}{a^2-\lambda_2}\right]+y_1^2\left[\frac{1}{b^2-\lambda_1}-\frac{1}{b^2-\lambda_2}\right]+z_1^2\left[\frac{1}{c^2-\lambda_1}-\frac{1}{c^2-\lambda_2}\right]=0,$$

which shows that $\alpha=90°$. In words:

The quadrics of a confocal system form a triply orthogonal system (Fig. 2–35).

When we express x, y, z in terms of the $\lambda_1, \lambda_2, \lambda_3$ determining the three orthogonal quadrics passing through the point (x, y, z):

$$\mathbf{x}=\mathbf{x}(\lambda_1, \lambda_2, \lambda_3), \tag{11–6}$$

we obtain a system of *elliptic coordinates* in space.

We thus have established the existence of other triply orthogonal systems than the trivial ones determined by Eqs. (11–2). This lends interest to the following theorem, also due to Dupin:

The surfaces of a triply orthogonal system intersect in the lines of curvature.

To prove it let us define the system by means of Eq. (11–1) and Eq. (11–4). From Eq. (11–4) we find by differentiation

$$\mathbf{x}_{uw}\cdot\mathbf{x}_v=-\mathbf{x}_u\cdot\mathbf{x}_{vw}=+\mathbf{x}_w\cdot\mathbf{x}_{uv}=-\mathbf{x}_v\cdot\mathbf{x}_{uw},$$

hence

$$\mathbf{x}_{vw}\cdot\mathbf{x}_u=\mathbf{x}_{wu}\cdot\mathbf{x}_v=\mathbf{x}_{uv}\cdot\mathbf{x}_w=0. \tag{11–7}$$

Let us single out for consideration the surface $w=0$. Its surface normal is in the direction of $\mathbf{x}_w$; the vectors $\mathbf{x}_u$ and $\mathbf{x}_v$ are tangent vectors, functions of the curvilinear coordinates (u, v) on the surface.

The equation $\mathbf{x}_u\cdot\mathbf{x}_v=0$ means that the F on this surface vanishes; the equation $\mathbf{x}_w\cdot\mathbf{x}_{uv}=0$ means, according to Eq. (5–9) that f vanishes. Coordinate systems for which $f=F=0$, are formed by lines of curvature. Instead of $w=$ constant we could take any surface of the set and reason in a similar way. The lines of intersection are therefore all lines of curvature.

This theorem does not give us much valuable information for the cases (11–2), but it gives us a simple way of finding the lines of curvature on an ellipsoid, a hyperboloid of one sheet, or a hyperboloid of two sheets. If, for instance, we wish to find the lines of curvature on the ellipsoid

$$\frac{x^2}{a^2}+\frac{y^2}{b^2}+\frac{z^2}{c^2}=1, \tag{11–8}$$

we "imbed" this ellipsoid in the triply orthogonal system (11–5). Then we see immediately:

The lines of curvature on the ellipsoid (11–8) are the curves in which the ellipsoid is intersected by the two sets of hyperboloids in the triply orthogonal system (11–5) to which the ellipsoid belongs.

The lines of curvature on the ellipsoid are therefore space curves of the fourth degree.

We have been able to see immediately that an ellipsoid can be "imbedded" in a triply orthogonal system, and so can the system of ellipsoids (Eq. (11–5), $\lambda < a^2$) to which it belongs. It can be shown, however, that not every single infinity of surfaces can thus be made part of a triply orthogonal system. For instance, a system of quadrics (11–8), in which a^2, b^2, c^2 depend on a parameter u, can be part of a triply orthogonal system only if

$$a^3(b^2 - c^2)a' + b^3(c^2 - a^2)b' + c^3(a^2 - b^2)c' = 0,^* \quad a' = da/du, \text{ etc.}$$

This is not satisfied, for instance, for a family of ellipsoids which are similar with respect to their common center. It can be shown that *a single infinity of surfaces must satisfy a partial differential equation of the third order to be "imbeddable" in a triply orthogonal system.*

Several textbooks (Forsyth, Eisenhart, et al.) have chapters on triply orthogonal surfaces. A full treatment is given in G. Darboux, *Leçons sur les systèmes orthogonaux et les coordonnées curvilignes* (Paris, 1910). *Gaston Darboux* (1842–1917), to whose works we must often refer, in 1880 succeeded Chasles as professor of higher geometry at the Paris Sorbonne and in 1900 succeeded Bertrand as secrétaire perpétuel of the Académie des Sciences. In his many papers and books he combined geometrical intuition with a mastery of algebra and analysis. His *Leçons* are not only a great source of information on surface theory, but also belong to the best written mathematical books of the nineteenth century.

EXERCISES

1. Show that the coordinate lines of the surfaces

$$\mathbf{y} = \mathbf{x}_1(u) + \mathbf{x}_2(v),$$

where $\mathbf{x}_1(u)$ and $\mathbf{x}_2(v)$ are arbitrary vector functions, are conjugate lines. These surfaces are called *translation surfaces*.

2. Find the family of curves on a right conoid conjugate to the orthogonal trajectories of the generating lines.

3. If two surfaces intersect each other along a curve at a constant angle, and if the curve is a line of curvature on one surface, then it is also a line of curvature on the other. Also prove the converse theorem (O. Bonnet, *Journ. Ec. Polyt.* **35**, 1853, a special case by Joachimsthal, *Journal für Mathem.* **30**, 1846, pp. 347–350).

4. *Third Fundamental form.* Apart from $\mathrm{I} = d\mathbf{x} \cdot d\mathbf{x}$ and $\mathrm{II} = -d\mathbf{x} \cdot d\mathbf{N}$ a third fundamental form $\mathrm{III} = d\mathbf{N} \cdot d\mathbf{N}$ is occasionally introduced. Prove that if K is the total curvature and M the mean curvature

$$K\mathrm{I} - 2M\,\mathrm{II} + \mathrm{III} = 0.$$

---

* See e.g. A. R. Forsyth, *Differential Geometry*, p. 453.

5. *Theorem of Beltrami-Enneper.* The torsion of the asymptotic lines through a point of the surface is $\pm\sqrt{-K}$ (E. Beltrami, 1866, *Opere matem. I*, p. 301). Hint: Use the result of Exercise 4.

6. *Spherical image.* When through a point unit vectors are drawn parallel to the surface normals along a curve C on a surface, the end points describe the *spherical image* of C on the unit sphere. Show that III (Exercise 4) is the square of the element of arc of the spherical image. Also show that at corresponding points the spherical image (a) of the lines of curvature is parallel to the lines of curvature, (b) of the asymptotic lines is perpendicular to the asymptotic lines.

7. Show that the spherical image of a family of curves of a surface S is orthogonal to its conjugate family on S.

8. Show that the systems of surfaces

(a) $\qquad x^2 + y^2 + z^2 = u, \qquad y = vx, \qquad x^2 + y^2 = wz,$

(b) $\quad x^2 + y^2 + z^2 = ux, \qquad x^2 + y^2 + z^2 = vy, \qquad x^2 + y^2 + z^2 = wz,$

form a triply orthogonal set.

9. Prove that the conics of the confocal system in the plane

$$\frac{x^2}{a^2 - \lambda} + \frac{y^2}{b^2 - \lambda} = 1 \qquad (\lambda \text{ variable})$$

form an orthogonal set (Fig. 2–36): (a) by following a method similar to that of Section (2–11), (b) by considering the transformation $w = \sin z$ of one complex plane on the other.

10. Prove that Eq. (11–5) in λ, with fixed x, y, z, has three real roots. Hint: Show that the roots of Eq. (11–5) are separated by the roots of the equation of Exercise 9.

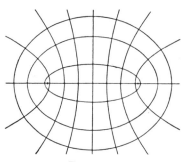

Fɪɢ. 2–36

CHAPTER 3

THE FUNDAMENTAL EQUATIONS

3-1 Gauss. The material of Chapters 1 and 2 bears the strong imprint of the methods of Monge. In this chapter and in the next we discuss another approach, connected with the name of the great German mathematician, Gauss.

For both men, theory and practice were intimately related. Monge, as a craftsman and engineer, saw in a surface primarily the boundary of a solid body, and consequently stressed the properties of a surface in relation to the surrounding space. He selected his topics, in the main, by their visual or engineering appeal, often finding his theoretical guidance in their possible relationship to the theory of partial differential equations. Gauss approached the theory of surfaces primarily as a result of his work on triangulation, where the emphasis is on measurements between points on the surface of the earth. Consequently he saw in a surface not so much the boundary of a solid body, as a fleece or film, a two-dimensional entity not necessarily attached to a three-dimensional body. A piece of such a surface can be bent and we can ask for the properties of the fleece which do not change under bending. A two-dimensional being, living on this surface and unaware of any outside space — like the beings of Abbott's *Flatland*,* which live in the plane unaware of any space of which the plane may be a part — would not be able to find out what asymptotic lines or lines of curvature are. But he would be able to find the road of shortest distance between two points measured along the surface, or the angle of two directions on the surface, that is, the *intrinsic* properties of the surface. Thus, with his characteristic understanding of theory and practice alike, Gauss drew from his work as a surveyor the inspiration for his profound reappraisal of the general theory of surfaces.

Carl Friedrich Gauss (1777–1855) was director of the astronomical observatory at Göttingen from 1807 to his death. One of the greatest scientists of all times, his work combined the fertility and particular originality of the eighteenth century mathematicians with the critical spirit of modern times.

The fields of his activity ranged from the theory of numbers to complex variable theory, and from celestial mechanics to the electric telegraph, of

---

* E. A. Abbott, *Flatland — a romance of many dimensions — by a square* (1884). Several editions; among others, Boston, Little, Brown & Co., 1941, xiii + 155 pp.

which he is one of the inventors. His fundamental work on surface theory, the *Disquisitiones generales circa superficies curvas* of 1827 (*Werke* 4, pp. 217–258) was, like much of Gauss' work, written in Latin, but there exists an English translation, *General investigations on curved surfaces,* Princeton, 1902, 127 pages. Gauss published other works on questions pertaining to geometry, such as his paper on conformal representation (1822), but he kept some of his boldest ideas to himself, notably his ideas on non-Euclidean geometry (see Section 4–7).

3–2 The equations of Gauss-Weingarten. We can penetrate into Gauss' mode of thinking by asking whether there exist any relations between the coefficients of the first and of the second fundamental form. It is clear that such relations cannot be purely algebraic, since E, F, G depend on only $\mathbf{x}_u$ and $\mathbf{x}_v$, whereas e, f, g depend also on $\mathbf{x}_{uu}$, $\mathbf{x}_{uv}$, and $\mathbf{x}_{vv}$ (algebraic relations can, of course, exist in special cases to characterize special surfaces or the nature of the surface at certain points, such as umbilics). The general relations between E, F, G, e, f, g must be differential relations. We shall now show how these relations can be found.

This demonstration can be accomplished by means of certain formulas which were used by Gauss himself in his paper of 1827. We shall derive them by introducing, at a point P of a surface, a *moving trihedron,* now composed, not of three mutually orthogonal unit vectors, as in the case of space curves, but of the three linearly independent vectors $\mathbf{x}_u$, $\mathbf{x}_v$, $\mathbf{N}$, of which $\mathbf{x}_u$ and $\mathbf{x}_v$ lie in the tangent plane normal to $\mathbf{N}$ (Fig. 3–1). These vectors satisfy the equations for scalar multiplication:

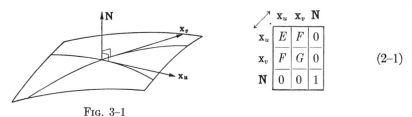

$\nearrow$	$\mathbf{x}_u$	$\mathbf{x}_v$	$\mathbf{N}$
$\mathbf{x}_u$	E	F	0
$\mathbf{x}_v$	F	G	0
$\mathbf{N}$	0	0	1

$$(2–1)$$

Fig. 3–1

This moving trihedron depends on two parameters (u, v), whereas in the case of space curves it depends on only one parameter.

Every vector can be linearly expressed in the three basic vectors of the moving trihedron. When we do this for $\mathbf{x}_{uu}$, $\mathbf{x}_{uv}$, and $\mathbf{x}_{vv}$, we obtain equations of the form

$$\begin{aligned}
\mathbf{x}_{uu} &= \alpha_1\mathbf{x}_u + \alpha_2\mathbf{x}_v + \alpha_3\mathbf{N}, \\
\mathbf{x}_{uv} &= \beta_1\mathbf{x}_u + \beta_2\mathbf{x}_v + \beta_3\mathbf{N}, \\
\mathbf{x}_{vv} &= \gamma_1\mathbf{x}_u + \gamma_2\mathbf{x}_v + \gamma_3\mathbf{N},
\end{aligned} \qquad (2–2)$$

where the coefficients $\alpha_1 \ldots \gamma_3$ have to be determined. We immediately find, by using Table 2–1 and Section 2–5, that

$$\alpha_3 = \mathbf{x}_{uu} \cdot \mathbf{N} = e, \quad \beta_3 = \mathbf{x}_{uv} \cdot \mathbf{N} = f, \quad \gamma_3 = \mathbf{x}_{vv} \cdot \mathbf{N} = g. \quad (2\text{–}3)$$

Moreover, using again Table 2–1 and introducing the notation

$$[11, 1] = \mathbf{x}_{uu} \cdot \mathbf{x}_u, \qquad [11, 2] = \mathbf{x}_{uu} \cdot \mathbf{x}_v, \quad (2\text{–}4)$$

we find

$$[11, 1] = E\alpha_1 + F\alpha_2, \qquad [11, 2] = F\alpha_1 + G\alpha_2,$$

or, solving these equations for α_1 and α_2,

$$\alpha_1 = \frac{G[11, 1] - F[11, 2]}{EG - F^2}, \quad \alpha_2 = \frac{E[11, 2] - F[11, 1]}{EG - F^2}. \quad (2\text{–}5)$$

The bracket symbols $[11, 1]$ and $[11, 2]$ can be expressed in terms of derivatives of E, F, G, because

$$2\mathbf{x}_u \cdot \mathbf{x}_{uu} = E_u, \qquad 2\mathbf{x}_u \cdot \mathbf{x}_{uv} = E_v, \qquad \mathbf{x}_u \cdot \mathbf{x}_{vu} + \mathbf{x}_{uu} \cdot \mathbf{x}_v = F_u.$$

Hence:

$$[11, 1] = \tfrac{1}{2}E_u, \qquad [11, 2] = F_u - \tfrac{1}{2}E_v,$$

and, if we introduce these expressions into Eq. (2–5), we see that the coefficients α_1 and α_2 can be expressed in terms of E, F, G and their first derivatives. A similar reasoning gives us β_1, β_2 and γ_1, γ_2.

Changing the notation, we denote α_1 by Γ^1_{11} and α_2 by Γ^2_{11}. If we change the β_i and $\gamma_i (i = 1, 2)$ similarly, we obtain the formulas

$$\begin{aligned}
\mathbf{x}_{uu} &= \Gamma^1_{11}\mathbf{x}_u + \Gamma^2_{11}\mathbf{x}_v + e\mathbf{N}, \\
\mathbf{x}_{uv} &= \Gamma^1_{12}\mathbf{x}_u + \Gamma^2_{12}\mathbf{x}_v + f\mathbf{N}, \\
\mathbf{x}_{vv} &= \Gamma^1_{22}\mathbf{x}_u + \Gamma^2_{22}\mathbf{x}_v + g\mathbf{N},
\end{aligned} \quad (2\text{–}6)$$

where the Γ^i_{jk} $(i, j, k = 1, 2)$ are defined as follows:

$$\begin{aligned}
\Gamma^1_{11} &= \frac{GE_u - 2FF_u + FE_v}{2(EG - F^2)}, & \Gamma^2_{11} &= \frac{2EF_u - EE_v - FE_u}{2(EG - F^2)} \\
\Gamma^1_{12} &= \frac{GE_v - FG_u}{2(EG - F^2)}, & \Gamma^2_{12} &= \frac{EG_u - FE_v}{2(EG - F^2)}, \\
\Gamma^1_{22} &= \frac{2GF_v - GG_u - FG_v}{2(EG - F^2)}, & \Gamma^2_{22} &= \frac{EG_v - 2FF_v + FG_u}{2(EG - F^2)}.
\end{aligned} \quad (2\text{–}7)$$

The Γ^i_{jk} are called the *Christoffel symbols*. It is convenient to introduce not only Γ^i_{jk} but also Γ^i_{kj} by the definition

$$\Gamma^i_{jk} = \Gamma^i_{kj} \quad (\text{hence } \Gamma^1_{21} = \Gamma^1_{12}, \quad \Gamma^2_{21} = \Gamma^2_{12}). \quad (2\text{–}8)$$

The $[i\,j,\,k]$ are often called *Christoffel symbols of the first kind;* in that case we call the Γ^i_{jk} *Christoffel symbols of the second kind.*   *They depend exclusively on the coefficients of the first fundamental form and their first derivatives.*

Eqs. (2–6) are the *Gauss equations* which we intend to derive.   They have also been called *the partial differential equations of surface theory.*

The Christoffel symbols are called after Erwin Bruno Christoffel (1829–1901), who taught first at Zürich and after the Franco-Prussian war of 1870–1871 at Strassbourg.   Christoffel introduced his symbols in a paper on differential forms in n variables, published in Crelle's *Journal für Mathem.* **70**, 1869 (*Gesamm. Abhandlungen I*, 1910, p. 352), denoting our Γ^i_{jk} by $\{^{jk}_i\}$.   The change to our present notation has been made under the influence of tensor theory.

It will be found useful to complement the Gauss Eqs. (2–6) by the two equations which express the derivatives $\mathbf{N}_u$ and $\mathbf{N}_v$ in terms of the moving trihedron.   Since $\mathbf{N}_u$ and $\mathbf{N}_v$ lie in the tangent plane, the expressions will be of the form

$$\mathbf{N}_u = p_1\mathbf{x}_u + p_2\mathbf{x}_v, \qquad \mathbf{N}_v = q_1\mathbf{x}_u + q_2\mathbf{x}_v.$$

With the aid of Table 2–1 and Section 2–5, we obtain

$$-e = p_1E + p_2F, \qquad -f = q_1E + q_2F,$$
$$-f = p_1F + p_2G, \qquad -g = q_1F + q_2G,$$

or

$$\mathbf{N}_u = \frac{fF - eG}{EG - F^2}\,\mathbf{x}_u + \frac{eF - fE}{EG - F^2}\,\mathbf{x}_v,$$
$$\mathbf{N}_v = \frac{gF - fG}{EG - F^2}\,\mathbf{x}_u + \frac{fF - gE}{EG - F^2}\,\mathbf{x}_v. \tag{2–9}$$

It is customary to call these equations after *J. Weingarten* (Crelle's *Journal für Mathem.* **59**, 1861).

An interesting application of the Gauss equations is obtained by referring the surface to a conjugate set of curvilinear coordinates.   Then $f = 0$, and the second of the Eqs. (2–6) becomes

$$\mathbf{x}_{uv} - \Gamma^1_{12}\mathbf{x}_u - \Gamma^2_{12}\mathbf{x}_v = 0. \tag{2–10}$$

This is a set of three differential equations for the rectangular coordinates x_i of the surface.   Conversely, when the second of the Eqs. (2–6) takes the form (2–10), the curvilinear coordinates (u, v) on the surface are conjugate. The Γ^1_{12}, Γ^2_{12} may be any functions of u and v, since their expressions in terms of the E, F, G and their derivatives of the corresponding surface are determined by Eq. (2–10) itself.   We thus have found the theorem:

Three functionally independent solutions of the partial differential equation

$$\varphi_{uv} - A\varphi_u - B\varphi_v = 0, \qquad A = A(u, v), \qquad B = B(u, v) \quad (2\text{–}11)$$

determine a surface on which the curves u = constant, v = constant are a conjugate set of curves.

When the curves v = constant are asymptotic, $e = 0$, and we find in a similar way:

Three functionally independent solutions of the partial differential equation

$$\varphi_{uu} - A\varphi_u - B\varphi_v = 0, \qquad A = A(u, v), \qquad B = B(u, v) \quad (2\text{–}12)$$

determine a surface on which the curves v = constant are asymptotic lines. These conditions are necessary and sufficient.

The importance of these theorems lies in the fact that the more general linear partial differential equation of the second order,

$$A\varphi_{uu} + B\varphi_{uv} + C\varphi_{vv} + D\varphi_u + E\varphi_v = 0,$$

where $A, \ldots E$ are functions of u and v, can always be transformed into either form (2–11) or form (2–12) by a real or imaginary substitution of the independent variables. For further information we refer to Darboux' *Surfaces* I, pp. 102–145, where many applications are made connecting the theory of partial differential equations, and their theory of characteristics, with that of surfaces. We confine ourselves to a very simple special case, obtained by taking $A = 0, B = 0$ in Eq. (2–11). Then we obtain

$$\varphi_{uv} = 0, \quad (2\text{–}13)$$

of which the general solution is $\varphi = f_1(u) + f_2(v)$, where f_1 and f_2 are arbitrary functions. Hence:

The surfaces

$$x_i = U_i(u) + V_i(v), i = 1, 2, 3, \quad (2\text{–}14)$$

where the U_i and V_i are arbitrary functions of u and v respectively, have a conjugate set of parametric lines. These surfaces are called *translation surfaces* (see Exercise 1, Section 2–11), since they can be obtained by moving the space curve $\mathbf{x} = \mathbf{U}(u)$ parallel to itself in such a way that one point of the curve moves along the space curve $\mathbf{x} = \mathbf{V}(v)$. The curves $\mathbf{U}(u)$ and $\mathbf{V}(v)$

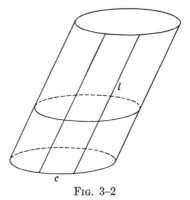

Fig. 3–2

may be interchanged and still yield the same surface. An example is a cylinder obtained by moving a straight line l parallel to itself along a circle c (Fig. 3–2); it can also be obtained by moving the circle c parallel to itself along the line l (that is, moving the plane of circle c with c in it parallel to itself such that c continues to intersect l). Straight lines and circles form a conjugate set of curves.

3–3 The theorem of Gauss and the equations of Codazzi. The equations of Gauss define the coordinates x_i of a surface as functions of u and v by means of a set of differential equations. These equations are not independent, but certain compatibility conditions are satisfied. They are expressed by the equations:

$$(\mathbf{x}_{uu})_v = (\mathbf{x}_{uv})_u; \qquad (\mathbf{x}_{uv})_v = (\mathbf{x}_{vv})_u.$$

We have therefore for any surface:

$$\frac{\partial}{\partial v}\left(\Gamma_{11}^1 \mathbf{x}_u + \Gamma_{11}^2 \mathbf{x}_v + e\mathbf{N}\right) = \frac{\partial}{\partial u}\left(\Gamma_{12}^1 \mathbf{x}_u + \Gamma_{12}^2 \mathbf{x}_v + f\mathbf{N}\right),$$

$$\frac{\partial}{\partial v}\left(\Gamma_{12}^1 \mathbf{x}_u + \Gamma_{12}^2 \mathbf{x}_v + f\mathbf{N}\right) = \frac{\partial}{\partial u}\left(\Gamma_{22}^1 \mathbf{x}_u + \Gamma_{22}^2 \mathbf{x}_v + g\mathbf{N}\right). \tag{3–1}$$

The terms in $\mathbf{x}_{uu}$, $\mathbf{x}_{uv}$, and $\mathbf{x}_{vv}$ can again be expressed in terms of $\mathbf{x}_u$, $\mathbf{x}_v$, and $\mathbf{N}$, by means of the Gauss equations, while the terms in $\mathbf{N}_u$ and $\mathbf{N}_v$ can be expressed in terms of $\mathbf{x}_u$ and $\mathbf{x}_v$ by means of the Weingarten equation. In this way we obtain two identities between the vectors $\mathbf{x}_u$, $\mathbf{x}_v$, $\mathbf{N}$, which can be satisfied only if the coefficients of $\mathbf{x}_u$, $\mathbf{x}_v$, and $\mathbf{N}$ are identically zero. In this way we obtain six scalar equations. We shall study in detail the three resulting from the first equation of (3–1):

(a) coeff. of $\mathbf{x}_u$: $\dfrac{\partial}{\partial v}\Gamma_{11}^1 + \Gamma_{11}^1\Gamma_{12}^1 + \Gamma_{11}^2\Gamma_{22}^1 + e\,\dfrac{gF - fG}{EG - F^2}$

$$= \frac{\partial}{\partial u}\Gamma_{12}^1 + \Gamma_{12}^1\Gamma_{11}^1 + \Gamma_{12}^2\Gamma_{12}^1 + f\,\frac{fF - eG}{EG - F^2}.$$

(b) coeff. of $\mathbf{x}_v$: $\dfrac{\partial}{\partial v}\Gamma_{11}^2 + \Gamma_{11}^1\Gamma_{12}^2 + \Gamma_{11}^2\Gamma_{22}^2 + e\,\dfrac{fF - gE}{EG - F^2}$ $\qquad$ (3–2)

$$= \frac{\partial}{\partial u}\Gamma_{12}^2 + \Gamma_{12}^1\Gamma_{11}^2 + \Gamma_{12}^2\Gamma_{12}^2 + f\,\frac{eF - fE}{EG - F^2}.$$

(c) coeff. of $\mathbf{N}$: $f\Gamma_{11}^1 + g\Gamma_{11}^2 + \dfrac{\partial e}{\partial v} = e\Gamma_{12}^1 + f\Gamma_{12}^2 + \dfrac{\partial f}{\partial u}.$

Three other equations can be derived from the second equation in (3–1). The Eqs. (3–2) can be written in the form:

(a) $\quad F\,\dfrac{eg - f^2}{EG - F^2} = \dfrac{\partial}{\partial u}\Gamma_{12}^1 - \dfrac{\partial}{\partial v}\Gamma_{11}^1 + \Gamma_{12}^2\Gamma_{12}^1 - \Gamma_{11}^2\Gamma_{22}^1.$

(b) $-E \dfrac{eg - f^2}{EG - F^2} = \dfrac{\partial}{\partial u} \Gamma^2_{12} - \dfrac{\partial}{\partial v} \Gamma^2_{11} + \Gamma^1_{12}\Gamma^2_{11} - \Gamma^1_{11}\Gamma^2_{12} + \Gamma^2_{12}\Gamma^2_{12} - \Gamma^2_{11}\Gamma^2_{22}.$

(c) $\qquad \dfrac{\partial e}{\partial v} - \dfrac{\partial f}{\partial u} = e\Gamma^1_{12} + f(\Gamma^2_{12} - \Gamma^1_{11}) - g\Gamma^2_{11}.$　　　　　　(3–3)

The Eqs. (3–3a) and (3–3b) have in common that they contain the coefficients of the second fundamental form only in the combination $eg - f^2$. Moreover, this combination appears as numerator of the fraction $(eg - f^2)/(EG - F^2)$, which is (see Section 2–7) the Gaussian curvature of the surface. Eqs. (3–3a) and (3–3b) both express the property that this Gaussian curvature depends only on E, F, G and their first and second derivatives. We call an expression which depends only on E, F, G and its derivatives a *bending invariant*, and thus have found Gauss' theorem:

The Gaussian curvature of a surface is a bending invariant. (This is a " *Theorema egregium*," "a most excellent theorem," wrote Gauss.)

This same result is expressed, in different analytical form, by the first two equations obtained from the second of Eqs. (3–1). The third equation gives us a result which has in common with the Eq. (3–3c) that it contains derivatives of coefficients of the second fundamental form. We write this new third equation together with (3–3c):

$$
\boxed{
\begin{aligned}
\dfrac{\partial e}{\partial v} - \dfrac{\partial f}{\partial u} &= e\Gamma^1_{12} + f(\Gamma^2_{12} - \Gamma^1_{11}) - g\Gamma^2_{11}, \\
\dfrac{\partial f}{\partial v} - \dfrac{\partial g}{\partial u} &= e\Gamma^1_{22} + f(\Gamma^2_{22} - \Gamma^1_{12}) - g\Gamma^2_{12}.
\end{aligned}
}
\qquad (3\text{–}4)
$$

These two equations are known as *the equations of Codazzi* or of *Mainardi-Codazzi*.

These formulas are found in the paper by *D. Codazzi* (1824–1875) in his answer to a "concours" of the Paris Academy (1860, printed in the *Mém. présentés à l'Académie* **27**, 1880); also in the *Annali di Matem.* **2**, 1868–1869, pp. 101–119. They were, at that time, already published by G. Mainardi, *Giornale Istituto Lombardo* **9**, 1856, pp. 385–398. Gauss, in his *Disquisitiones* has these formulas in principle, though not explicitly. Their fundamental importance was first fully recognized by O. Bonnet, *Journal de l'Ecole Polytechnique* **42** (1867), pp. 31–151.

There also exist compatibility conditions for the Weingarten equations (2–9). However, these equations, obtained by expressing that $\mathbf{N}_{uv} = \mathbf{N}_{vu}$ and by using the Gauss equations (2–6), can be shown to be equivalent to the Codazzi equations (see Exercise 21, Section 4–2).

When we call the Gaussian curvature a bending invariant, we mean that it is unchanged by such deformations of the surface (we always think of a limited region) which do not involve stretching, shrinking, or tearing. This *bending* leaves the distance between two points on the surface, measured along a curve on the surface, unchanged, and also the angle of two tangent directions at a point.   We can easily obtain an idea of this bending by deforming a piece of paper without changing its elastic properties (we should not wet it); a curve drawn on this piece of paper conserves its length even if its shape is changed as a result of the bending of the paper.

When the curvilinear coordinate lines retain their position on the surface during the bending and the measurement of the coordinates remains the same, then the coefficients of the first fundamental form and all their derivatives with respect to the curvilinear coordinates also remain the same.   *A function containing E, F, G and their derivatives is therefore a bending invariant; such an invariant can also contain arbitrary functions $\varphi(u, v)$, $\psi(u, v)$, . . . or the derivatives dv/du, d^2v/du^2, ...*   Properties of surfaces expressible by bending invariants are called *intrinsic properties.*

Both the expression of Gauss' theorem and Codazzi's formulas can be written in many forms.   The formulation (3–3) of Gauss' theorem does not give an expression for the Gaussian curvature in which E, F, G take equivalent positions.   F. Brioschi, in 1852 (*Opere* I, p. 1) gave an expression which is more satisfactory from this point of view.   To obtain it we use the determinant expressions for e, f, g (Section 2–5):

$$K = \frac{eg - f^2}{EG - F^2} = \frac{1}{(EG - F^2)^2} \left[(\mathbf{x}_{uu}\mathbf{x}_u\mathbf{x}_v)(\mathbf{x}_{vv}\mathbf{x}_u\mathbf{x}_v) - (\mathbf{x}_{uv}\mathbf{x}_u\mathbf{x}_v)^2 \right],$$

$$= \frac{1}{(EG-F^2)^2} \left\{ \begin{vmatrix} \mathbf{x}_{uu}\cdot\mathbf{x}_{vv} & \mathbf{x}_{uu}\cdot\mathbf{x}_u & \mathbf{x}_{uu}\cdot\mathbf{x}_v \\ \mathbf{x}_u\cdot\mathbf{x}_{vv} & \mathbf{x}_u\cdot\mathbf{x}_u & \mathbf{x}_u\cdot\mathbf{x}_v \\ \mathbf{x}_v\cdot\mathbf{x}_{vv} & \mathbf{x}_v\cdot\mathbf{x}_u & \mathbf{x}_v\cdot\mathbf{x}_v \end{vmatrix} - \begin{vmatrix} \mathbf{x}_{uv}\cdot\mathbf{x}_{uv} & \mathbf{x}_{uv}\cdot\mathbf{x}_u & \mathbf{x}_{uv}\cdot\mathbf{x}_v \\ \mathbf{x}_u\cdot\mathbf{x}_{uv} & \mathbf{x}_u\cdot\mathbf{x}_u & \mathbf{x}_u\cdot\mathbf{x}_v \\ \mathbf{x}_v\cdot\mathbf{x}_{uv} & \mathbf{x}_v\cdot\mathbf{x}_u & \mathbf{x}_v\cdot\mathbf{x}_v \end{vmatrix} \right\} \cdot \quad (3\text{–}5)$$

In the terms of the determinants we recognize E, F, G and the symbols $[ij,k]$ of Eq. (2–4).   We still need expressions for $\mathbf{x}_{uu}\cdot\mathbf{x}_{vv}$ and $\mathbf{x}_{uv}\cdot\mathbf{x}_{uv}$, but since each of these expressions occurs with the same factor $EG - F^2$, we need only their difference.   We can verify, by the same method as used for Eq. (2–4), that

$$\mathbf{x}_{uu} \cdot \mathbf{x}_{vv} - \mathbf{x}_{uv} \cdot \mathbf{x}_{uv} = -\tfrac{1}{2}E_{vv} + F_{uv} - \tfrac{1}{2}G_{uu}.$$

We thus obtain for K the expression

$$K = \frac{1}{(EG-F^2)^2} \left\{ \begin{vmatrix} -\tfrac{1}{2}E_{vv}+F_{uv}-\tfrac{1}{2}G_{uu} & \tfrac{1}{2}E_u & F_u-\tfrac{1}{2}E_v \\ F_v-\tfrac{1}{2}G_u & E & F \\ \tfrac{1}{2}G_v & F & G \end{vmatrix} - \begin{vmatrix} 0 & \tfrac{1}{2}E_v & \tfrac{1}{2}G_u \\ \tfrac{1}{2}E_v & E & F \\ \tfrac{1}{2}G_u & F & G \end{vmatrix} \right\} \cdot \quad (3\text{–}6)$$

This takes a much simpler form in the case that the parametric lines are orthogonal. Then $F = 0$ and we can verify without much difficulty that in this case

$$K = -\frac{1}{\sqrt{EG}}\left[\frac{\partial}{\partial u}\left(\frac{1}{\sqrt{E}}\frac{\partial\sqrt{G}}{\partial u}\right) + \frac{\partial}{\partial v}\left(\frac{1}{\sqrt{G}}\frac{\partial\sqrt{E}}{\partial v}\right)\right]. \tag{3-7}$$

Eqs. (3-5) through (3-7) are also called *Gauss' equations*. They all express the "Theorema egregium."

The Codazzi equations take a simple form when we do not only take $F = 0$, but also $f = 0$, which means that we take the lines of curvature as coordinate lines. Then (Sec. 2-6) we have for the principal normal curvatures κ_1 and κ_2:

$$\kappa_1 = e/E, \qquad \kappa_2 = g/G,$$

and the Codazzi equations, now reduced to

$$e_v = \tfrac{1}{2}E_v\left(\frac{e}{E} + \frac{g}{G}\right), \qquad g_u = \tfrac{1}{2}G_u\left(\frac{e}{E} + \frac{g}{G}\right), \tag{3-7a}$$

take the form

$$\frac{\partial\kappa_1}{\partial v} = \frac{1}{2}\frac{E_v}{E}(\kappa_2 - \kappa_1), \qquad \frac{\partial\kappa_2}{\partial u} = \frac{1}{2}\frac{G_u}{G}(\kappa_1 - \kappa_2). \tag{3-8}$$

EXERCISES

1. Derive the second of the Codazzi Eq. (3-4) from the second Eq. (3-1).

2. Derive Rodrigues' formula for the lines of curvature from the Weingarten equations.

3. Compute the Christoffel symbols for polar coordinates in the plane.

4. Show that when the surface is given by $z = f(x, y)$ (see Exercise 2, Sec. 2-8),

$$\Gamma_{11}^1 = \frac{pr}{1 + p^2 + q^2}, \qquad \Gamma_{12}^1 = \frac{ps}{1 + p^2 + q^2}, \qquad \Gamma_{22}^1 = \frac{pt}{1 + p^2 + q^2},$$

$$\Gamma_{11}^2 = \frac{qr}{1 + p^2 + q^2}, \quad \Gamma_{12}^2 = \frac{qs}{1 + p^2 + q^2}, \quad \Gamma_{22}^2 = \frac{qt}{1 + p^2 + q^2}, \quad K = \frac{rt - s^2}{(1 + p^2 + q^2)^2}.$$

5. The locus of the centers of the chords of a space curve C is a translation surface on which C is an asymptotic curve.

6. Show that the paraboloid $z = ax^2 + by^2$ is a translation surface. What are the curves $u = $ constant, $v = $ constant (Eq. 2-14) in this case?

7. When $D^2 = EG - F^2$, show that

$$\frac{\partial}{\partial u}\ln D = \Gamma_{11}^1 + \Gamma_{12}^2, \qquad \frac{\partial}{\partial v}\ln D = \Gamma_{12}^1 + \Gamma_{22}^2.$$

8. When ω is the angle between the coordinate curves, show that

$$\frac{\partial \omega}{\partial u} = -\frac{D}{E}\Gamma_{11}^2 - \frac{D}{G}\Gamma_{12}^1, \qquad \frac{\partial \omega}{\partial v} = -\frac{D}{E}\Gamma_{12}^2 - \frac{D}{G}\Gamma_{22}^1.$$

9. Verify the simplified equations of Gauss and of Codazzi (3–7) and (3–8).

10. Verify the following expressions for the Gaussian curvature:

$$K = -\frac{1}{4(EG-F^2)^2}\begin{vmatrix} E & F & G \\ E_u & F_u & G_u \\ E_v & F_v & G_v \end{vmatrix} - \frac{1}{2\sqrt{EG-F^2}}\left\{\frac{\partial}{\partial u}\frac{G_u-F_v}{\sqrt{EG-F^2}} - \frac{\partial}{\partial v}\frac{F_u-E_v}{\sqrt{EG-F^2}}\right\},$$

(G. Frobenius, see *Blaschke* **I**, p. 117.)

$$K = \frac{1}{D}\left[\frac{\partial}{\partial v}\left(\frac{D}{E}\Gamma_{11}^2\right) - \frac{\partial}{\partial u}\left(\frac{D}{E}\Gamma_{12}^2\right)\right] = \frac{1}{D}\left[\frac{\partial}{\partial u}\left(\frac{D}{G}\Gamma_{22}^1\right) - \frac{\partial}{\partial v}\left(\frac{D}{G}\Gamma_{12}^1\right)\right].$$

(J. Liouville, *Journal de Mathém.* **16** (1851), p. 130.)

11. Show that the direction cosines N_i of the unit surface normal vector satisfy equations of the form

$$\varphi_{uv} + A\varphi_u + B\varphi_v + C\varphi = 0,$$
$$\varphi_{uu} + A_1\varphi_u + B_1\varphi_v + C_1\varphi = 0, \qquad A \ldots C_1 \text{ functions of } u \text{ and } v.$$

12. Show that K is invariant under a change of curvilinear coordinates on the surface.

13. *Tensor notation.* We introduce the notation

$$g_{11} = E, \qquad\qquad g_{12} = F, \qquad\qquad g_{22} = G,$$
$$g^{11} = \frac{G}{EG - F^2}, \qquad g^{12} = \frac{-F}{EG - F^2}, \qquad g^{22} = \frac{E}{EG - F^2}.$$

Show that

$$g_{ij}g^{ik} = \delta_j^k, \qquad i, j, k = 1, 2.$$

Here *we sum on the index which is repeated* (here i). The *Kronecker symbol* δ_j^k is 1 when $k = j$, and zero when $k \neq j$; for example $\delta_1^1 = 1$, $\delta_2^1 = 0$.

In the next exercises we use the notation of Exercise 13; the indices $i, j, k, \ldots$ run over 1 and 2. We call $u = u_1$, $v = u_2$.

14. Show that

$$[ij, k] = \frac{1}{2}\left(\frac{\partial g_{ik}}{\partial u_j} + \frac{\partial g_{jk}}{\partial u_i} - \frac{\partial g_{ij}}{\partial u_k}\right),$$
$$\Gamma_{jk}^t = g^{il}[jk, l],$$
$$[jk, l] = g_{li}\Gamma_{jk}^t.$$

15. Show that Gauss' equations (2–6) can be written

$$\mathbf{x}_{ij} = \Gamma_{ij}^k\mathbf{x}_k + h_{ij}\mathbf{N},$$

$\mathbf{x}_1 = \mathbf{x}_u$, $\mathbf{x}_2 = \mathbf{x}_v$,   etc., and $e = h_{11}$, $f = h_{21} = h_{12}$, $g = h_{22}$.

16. Show that Weingarten's equations (2–9) can be written:

$$\mathbf{N}_i = -\,(g^{kl}h_{il})\mathbf{x}_k.$$

(Here the summing is both on l and on k.)

17. Show that Codazzi's equations can be written:

$$\frac{\partial h_{ij}}{\partial u_l} - \Gamma_{il}^k h_{kj} = \frac{\partial h_{il}}{\partial u_j} - \Gamma_{ij}^k h_{kl}.$$

18. By introducing the *symbol of Riemann:*

$$\{kl,\,ij\} = R_{ijk}^l = \frac{\partial}{\partial u_j}\,\Gamma_{ik}^l - \frac{\partial}{\partial u_i}\,\Gamma_{jk}^l + \Gamma_{ik}^m \Gamma_{mj}^l - \Gamma_{jk}^m \Gamma_{mi}^l \text{ (sum on } m),$$

show that the Eqs. (3–3) can be written as

$$g^{n2}(eg - f^2) = R_{121}^n = \{1n,\,12\}.$$

19. By introducing the symbol

$$(kl,\,ij) = R_{ijkl} = R_{ijk}^m g_{ml},$$

show that the Gaussian curvature K can be written:

$$K = \frac{R_{1212}}{EG - F^2} = \frac{(12,\,12)}{EG - F^2}.$$

20. Show that when we substitute E, F, G for e, f, g into the Codazzi equations, these equations are identically satisfied.

3–4 Curvilinear coordinates in space. The theory developed in the two previous sections can be considered as an aspect of a more general theory, dealing with families of surfaces in space. We shall give an outline of this theory for the special case that our given surface belongs to a triply orthogonal system of surfaces, which for one surface does not involve a restriction of generality. In this case we shall take the parametric lines along the curves of intersection of the surfaces. This means not only that the parametric lines are orthogonal, but also, according to Dupin's theorem, that they are lines of curvature. Using the notation of Section 2–11, we introduce the curvilinear coordinates (u, v, w) by means of the equation

$$\mathbf{x} = \mathbf{x}(u, v, w),$$

which establishes three families of ∞^1 surfaces

$$u(x, y, z) = c_1, \quad v(x, y, z) = c_2, \quad w(x, y, z) = c_3,$$

or for short:

$$u_i = c_i, \qquad\qquad i = 1, 2, 3. \qquad (4\text{–}1)$$

We also know (Section 2–11) that the relations exist:

$$\mathbf{x}_u \cdot \mathbf{x}_v = \mathbf{x}_v \cdot \mathbf{x}_w = \mathbf{x}_w \cdot \mathbf{x}_u = 0, \qquad\qquad (4\text{–}2\text{a})$$

$$\mathbf{x}_{uv} \cdot \mathbf{x}_w = \mathbf{x}_{vw} \cdot \mathbf{x}_u = \mathbf{x}_{wu} \cdot \mathbf{x}_v = 0, \qquad\qquad (4\text{–}2\text{b})$$

of which the last three express Dupin's theorem.   We can consider the surface of the previous sections as one of the surfaces $w = u_3 =$ constant. For these surfaces $f = 0$, $F = 0$.

At each point we have a set of three linearly independent orthogonal vectors $\mathbf{x}_u$, $\mathbf{x}_v$, $\mathbf{x}_w$, and also the set of the gradient vectors ∇u, ∇v, ∇w, which have the same directions as the $\mathbf{x}_u$, $\mathbf{x}_v$, $\mathbf{x}_w$ respectively.

The gradient vector field of a function of position $f(x, y, z)$ is defined as

$$\nabla f = \operatorname{grad} f = \mathbf{e}_1 f_x + \mathbf{e}_2 f_y + \mathbf{e}_3 f_z.$$

Since $df = d\mathbf{x} \cdot \nabla f$ this vector field is perpendicular to the surface $f =$ constant. The length of ∇f is df/dn, where dn is the element of length in the direction normal to $f =$ constant.

The ds^2 of space, expressed in the curvilinear coordinates u, v, w, takes the form (Section 2–11):

$$ds^2 = (\mathbf{x}_u \cdot \mathbf{x}_u) \, du^2 + (\mathbf{x}_v \cdot \mathbf{x}_v) \, dv^2 + (\mathbf{x}_w \cdot \mathbf{x}_w) \, dw^2,$$

for which it is customary to write either one of the following notations:

$$ds^2 = g_{11} \, du^2 + g_{22} \, dv^2 + g_{33} \, dw^2 = H_1^2 \, du^2 + H_2^2 \, dv^2 + H_3^2 \, dw^2. \quad (4\text{–}3)$$

The g-notation is called the *tensor notation* (see Exercise 13, Section 3–3), the H-notation is called *Lamé's notation* for this ds^2. We take the H_i positive, H_i^2 means $H_i H_i = (H_i)^2$.

If we now introduce three unit vectors $\mathbf{u}_1$, $\mathbf{u}_2$, $\mathbf{u}_3$ in the direction of $\mathbf{x}_u$, $\mathbf{x}_v$, $\mathbf{x}_w$ respectively (in the sense of increasing u, v, w), then since

$$|\nabla u| = du/dn = H_1^{-1}, \qquad |\nabla u_i| = H_i^{-1}, \qquad i = 1, 2, 3,$$

the following relations exist at each point P of space:

$$\mathbf{u}_1 = h_1 \mathbf{x}_u = H_1 \nabla u, \quad \mathbf{u}_2 = h_2 \mathbf{x}_v = H_2 \nabla v, \quad \mathbf{u}_3 = h_3 \mathbf{x}_w = H_3 \nabla w, \quad (4\text{–}4)$$

where the h_i are defined by

$$h_i = H_i^{-1}, \qquad\qquad\qquad i = 1, 2, 3.$$

The trihedrons $(\mathbf{x}_u, \mathbf{x}_v, \mathbf{x}_w)$, $(\mathbf{u}_1, \mathbf{u}_2, \mathbf{u}_3)$, and $(\nabla u, \nabla v, \nabla w)$ depend on the three parameters (u, v, w), and their motion in space can be used to explore the geometry of space very much in the way we have used the moving trihedron depending on one parameter for the investigation of space curves. This motion can be expressed in terms of the derivatives of either one of these trihedrons.   Following the method of Section 3–2, we write in analogy to the Gauss equations:

$$\begin{aligned} \mathbf{x}_{uu} &= \Gamma_{11}^1 \mathbf{x}_u + \Gamma_{11}^2 \mathbf{x}_v + \Gamma_{11}^3 \mathbf{x}_w \qquad \text{(and 2 similar equations)}, \\ \mathbf{x}_{uv} &= \Gamma_{12}^1 \mathbf{x}_u + \Gamma_{12}^2 \mathbf{x}_v + \Gamma_{12}^3 \mathbf{x}_w \qquad \text{(and 2 similar equations)}. \end{aligned} \quad (4\text{–}5)$$

Eqs. (4–2b) show that all Γ^i_{jk} $(i, j, k = 1, 2, 3,$ all different) vanish. For the other Γ we find by a method similar to that used in Section 3–2:

$$\Gamma^1_{11} = h_1\,\partial_u H_1 = -H_1\,\partial_u h_1 = \frac{dH_1}{ds_1}\ (=E_u/2E \text{ for } w = c),$$

$$\Gamma^1_{12} = h_1\,\partial_v H_1 = -H_1\,\partial_v h_1 = h_1 H_2 \frac{dH_1}{ds_2}\ (=E_v/2E \text{ for } w = c), \quad (4\text{–}6)$$

$$\Gamma^2_{11} = -H_1 h_2^2\,\partial_v H_1 = -H_1 h_2 \frac{dH_1}{ds_2}\ (=G_u/2E \text{ for } w = c),$$

and in general:

$$\Gamma^i_{ii} = \frac{dH_i}{ds_i}, \qquad \Gamma^i_{ij} = h_i H_j \frac{dH_i}{ds_j}, \qquad \Gamma^j_{ii} = -H_j h_j \frac{dH_i}{ds_j},*$$

where $ds_i = H_i\,du_i$ represents the ds in the direction normal to $u_i = $ constant. Hence

$$\mathbf{x}_{uu} = \frac{dH_1}{ds_1}\mathbf{x}_u - h_2 H_1 \frac{dH_1}{ds_2}\mathbf{x}_v - h_3 H_1 \frac{dH_1}{ds_3}\mathbf{x}_w, \text{ etc.}$$

$$\mathbf{x}_{uv} = h_1 H_2 \frac{dH_2}{ds_2}\mathbf{x}_u + h_2 H_1 \frac{dH_2}{ds_1}\mathbf{x}_v, \text{ etc.}$$
$$(4\text{–}7)$$

The corresponding equations for the unit vectors are obtained from equations such as $\partial_u\mathbf{u}_1 = \partial_u(h_1\mathbf{x}_u)$. We obtain:

$$\partial_u\mathbf{u}_1 = -\frac{dH_1}{ds_2}\mathbf{u}_2 - \frac{dH_1}{ds_3}\mathbf{u}_3,$$

$$\partial_v\mathbf{u}_1 = \frac{dH_2}{ds_1}\mathbf{u}_2,$$
$$(4\text{–}8)$$

$$\partial_w\mathbf{u}_1 = \frac{dH_3}{ds_1}\mathbf{u}_3, \text{ and 2 other sets of 3 equations.}$$

Since $\mathbf{u}_i$ is the unit normal vector of surface $u_i = $ const, the equations for $\partial_i\mathbf{u}_j$, $i \neq j$, express Rodrigues' theorem. The equation for $h_1\,\partial_u\mathbf{u}_1$ expresses the decomposition of the curvature vector of the lines of curvature in its tangential and normal component.

When two arbitrary functions $E(u, v)$, $G(u, v)$ of two variables are given, then it is always possible to find surfaces for which $ds^2 = E\,du^2 + G\,dv^2$ is the first fundamental form (see Section 5–3). This is not the case when three arbitrary functions $g_{11}(u, v, w)$, $g_{22}(u, v, w)$, $g_{33}(u, v, w)$ are given, and it is required that they be used as coefficients of a ds^2 of space of the form

---

* Contrary to Exercises 13–19, Section 3–3, we do not here sum on indices which occur more than once.

(4-3). It can be shown that the g_{ii} must satisfy six differential relations, the *compatibility equations*. We can obtain them, in a way similar to that used in Section 3-3, by evaluating equations of the type $\mathbf{x}_{uuv} = \mathbf{x}_{uvu}$. We obtain them in a simpler way by performing a similar operation on the unit vectors in Eq. (4-8):

$$\partial_u \, \partial_v \mathbf{u}_1 = \partial_v \, \partial_u \mathbf{u}_1, \text{ etc.}$$

We obtain

$$\partial_u \left[\frac{dH_2}{ds_1} \, \mathbf{u}_2 \right] = \partial_v \left[-\frac{dH_1}{ds_2} \, \mathbf{u}_2 - \frac{dH_1}{ds_3} \, \mathbf{u}_3 \right],$$

from which, by the use of Eq. (4-8), we obtain three scalar equations as coefficients of $\mathbf{u}_1, \mathbf{u}_2, \mathbf{u}_3$ respectively. The coefficient of $\mathbf{u}_1$ is identically zero. The other coefficients are

$$\frac{\partial}{\partial u} \frac{dH_2}{ds_1} + \frac{\partial}{\partial v} \frac{dH_1}{ds_2} = -\frac{dH_1}{ds_3} \frac{dH_2}{ds_3}, \tag{4-9}$$

$$\frac{\partial}{\partial v} \frac{dH_1}{ds_3} = \frac{dH_1}{ds_2} \frac{dH_2}{ds_3}. \tag{4-10}$$

There are three equations of the form (4-9), and six of the form (4-10). As we said before, it can be shown that six of them are independent, but we shall not prove this here (see Exercise 6, this section).   The Eq. (4-10) can also be written in the form

$$\frac{\partial^2 H_1}{\partial v \, \partial w} = \frac{dH_1}{ds_3} \frac{\partial H_3}{\partial v} + \frac{dH_1}{ds_2} \frac{\partial H_2}{\partial w} = (h_3 \partial_w H_1)(\partial_v H_3) + (h_2 \partial_v H_1)(\partial_w H_2). \tag{4-10a}$$

The equations of the type (4-9) and (4-10) are equivalent with the Gauss-Codazzi equations for the surfaces $u_i = $ constant. To show it for the surfaces $w = $ constant we notice that by comparison of Eqs. (4-5) and (2-6) we can express e and g in terms of Γ:

$$e = H_3 \Gamma_{11}^3 = -H_1 \frac{dH_1}{ds_3} = -\frac{1}{2} \frac{dg_{11}}{ds_3},$$
$$g = H_3 \Gamma_{22}^3 = -H_2 \frac{dH_2}{ds_3} = -\frac{1}{2} \frac{dg_{22}}{ds_3}. \tag{4-11}$$

We can write these equations also in the form $e = -\frac{1}{2} \, dE/ds_3$ and $g = -\frac{1}{2} \, dG/ds_3$, provided we understand that the E and G are here functions of three variables, in which $w = $ constant is substituted after differentiation. *They are interesting expressions for the coefficients of the second fundamental form in terms of those of the first.*

Eq. (4-9) becomes, after substituting for dH_1/ds_3 and dH_2/ds_3 their expressions (4-11) in terms of e and g,

$$\frac{\partial}{\partial u}(h_1 H_2 \Gamma_{12}^2) - \frac{\partial}{\partial v}(h_1 H_2 \Gamma_{11}^2) = h_1 h_2 eg,$$

which is equivalent with the Gauss equation (3–3b), in which $f = F = 0$.
Similarly, Eq. (4–10) becomes

$$\frac{\partial}{\partial v}(-h_1 c) = (h_2 H_1 \Gamma_{12}^1)(-h_2 g),$$

which is equivalent with the first of the Codazzi equations (3–4).

The equations of this section were derived by *Gabriel Lamé* (1795–1870), who in and after 1837 introduced curvilinear coordinates for the solution of the differential equations of heat and elasticity. He collected his results in the *Leçons sur les coordonnées curvilignes* (1859). He introduced the concepts of isothermic systems (see Section 5–2) and differential parameters (see Section 4–8, Exercise 11).

His work was used by G. Darboux for his investigations of triply orthogonal systems (Section 2–11); he generalized it to systems which intersect in conjugate systems (*Leçons sur les systèmes orthogonaux*, p. 361). Formulas for general systems of curvilinear coordinates in space can be found in papers by the Abbé Aoust (e.g. *Annali di matem.* **6**, 1864, pp. 65–87) who generalized Lamé's work. This work has since 1887 been superseded by that of G. Ricci-Curbastro, who created in the tensor calculus a new instrument to deal with questions pertaining to general curvilinear coordinates, not only for the geometry of three, but also for $n > 3$ dimensions.

EXERCISES

1. The sets of mutually orthogonal vectors $\mathbf{a}_i$, $\mathbf{b}_i$, $(i = 1, 2, 3)$, are called *reciprocal* if $\mathbf{a}_i \cdot \mathbf{b}_j = 0$, $i \neq j$; $\mathbf{a}_1 \cdot \mathbf{b}_1 = \mathbf{a}_2 \cdot \mathbf{b}_2 = \mathbf{a}_3 \cdot \mathbf{b}_3 = 1$. Show that when $\mathbf{x} = \mathbf{x}(u, v, w)$ introduces a set of curvilinear coordinates (not necessarily orthogonal), the sets $\mathbf{x}_u$, $\mathbf{x}_v$, $\mathbf{x}_w$ and ∇u, ∇v, ∇w form a reciprocal set.

2. Show that, in the notation of Section 3–4,

$$H_1^2 u_x^2 + H_2^2 u_y^2 + H_3^2 w_x^2 = 1, \qquad H_1^2 u_x u_y + H_2^2 v_x v_y + H_3^2 w_x w_y = 0, \text{ cycl.}$$
$$h_1^2 x_u^2 + h_2^2 x_v^2 + h_3^2 x_w^2 = 1, \qquad h_1^2 x_u y_u + h_2^2 x_v y_v + h_3^2 x_w y_w = 0, \text{ cycl.}$$

3. Show that $h_1 x_u = h_2 h_3 (y_v z_w - y_w z_v)$, cycl.,
$H_1 u_x = H_2 H_3 (v_y w_z - v_z w_y)$, cycl.

4. Find the rotation vector $\mathbf{p}$ which expresses the motion of the system of unit vectors $(\mathbf{u}_1, \mathbf{u}_2, \mathbf{u}_3)$ of Section 3–4 in the direction of $\mathbf{u}_1$.

5. Show that Eqs. (4–7) and (4–8) are equivalent with

$$\frac{\partial}{\partial u}\nabla u = -\Gamma_{11}^1 \nabla u - \Gamma_{12}^1 \nabla v - \Gamma_{13}^1 \nabla w, \text{ and 2 other equations,}$$

$$\frac{\partial}{\partial v}\nabla u = -\Gamma_{12}^1 \nabla u - \Gamma_{22}^1 \nabla v, \text{ and 5 other equations.}$$

6. Show that Eqs. (4–9) and (4–10) can be written (see Exercise 18, Section 3–3):

$$R^l_{ijk} = 0, \text{ where } i, j, k, l = 1, 2, 3.$$

(It can be proved that 6 of the R^l_{ijk} are independent, see L. P. Eisenhart, *Riemannian geometry*, Princeton, 1926, p. 21.)

3–5 Some applications of the Gauss and the Codazzi equations.
(1) An interesting illustration of Gauss' theorem is presented by the catenoid,

$$x = u \cos v, \qquad y = u \sin v, \qquad z = c \cosh^{-1} \frac{u}{c} = f(u), \qquad (5\text{–}1)$$

and the right helicoid,

$$x = u_1 \cos v_1, \qquad y = u_1 \sin v_1, \qquad z = a v_1 = f_1(v). \qquad (5\text{–}2)$$

The first fundamental form of the catenoid is (Section 2–8):

$$ds^2 = \frac{u^2}{u^2 - c^2} du^2 + u^2 \, dv^2,$$

that of the right helicoid,

$$ds_1^2 = du_1^2 + (u_1^2 + a^2) \, dv_1^2.$$

If we now write

$$a = c, \qquad v = v_1, \qquad u = \sqrt{u_1^2 + a^2} \quad \text{or} \quad u_1 = \sqrt{u^2 - a^2} \qquad (5\text{–}3)$$

then

$$ds_1^2 = ds^2,$$

and we see that for $0 \leqslant v < 2\pi$ and $-a \leqslant u < + a$ we have established a one-to-one correspondence between the points on both surfaces such that at corresponding points the first fundamental forms are equal.
The total curvature of the catenoid is (Section 2–8):

$$K = \frac{f'f''}{u(1 + f'^2)^2} = \frac{-c^2}{u^4}, \qquad \text{since } f' = \frac{c}{\sqrt{u^2 - c^2}}.$$

The total curvature of the right helicoid is:

$$K_1 = \frac{-(f_1')^2}{[u_1^2 + (f_1')^2]^2} = \frac{-a^2}{(u_1^2 + a^2)^2},$$

and we can readily verify that the substitution (5–3) also leads to $K = K_1$. This shows that *a correspondence can be established between the points of a catenoid and of a right helicoid such that at corresponding points the E, F, G and therefore the Gaussian curvature are the same.*

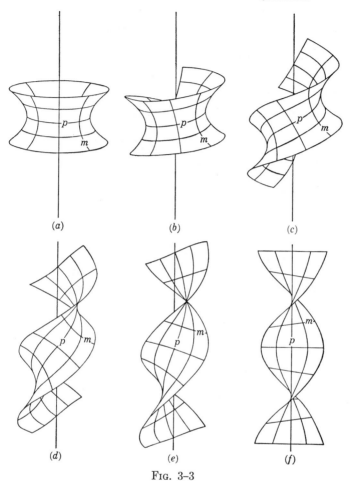

(a)   (b)   (c)

(d)   (e)   (f)

FIG. 3-3

This is a one-to-one correspondence so long as $-a \leqslant u < a$, and $0 \leqslant v < 2\pi$, hence for one full turn of the helix.

It can be shown (Section 5-4) that one surface can actually pass into the other by a continuous bending. In Fig. 3-3 we show six different stages in this deformation. This can be demonstrated with a flexible piece of brass applied to a plaster model of a catenoid and bent so as to be applied to a model of a right helicoid (Fig. 3-4, models by Brill, Darmstadt). Eqs. (5-2) and (5-3) show that the circles on the catenoid pass into the helices of the helicoid, and the catenaries of the catenoid into the straight lines of

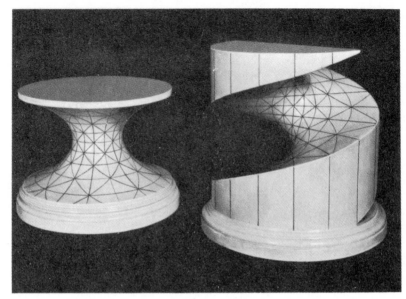

Fig. 3–4

the helicoid (that is, for the parts of the surfaces for which $-a \leqslant u < +a$, $0 \leqslant v < 2\pi$, hence for one full turn of the helix).

(2) As an illustration of the Codazzi equations we prove the theorem:

The sphere is the only surface all points of which are umbilics.

The normal curvatures at an umbilic are all equal, and we can write $\kappa = \kappa_1 = \kappa_2$ (Section 2–7). The Codazzi Eqs. (3–8) then show that $\partial \kappa / \partial u = 0$ and $\partial \kappa / \partial v = 0$, hence that κ is a constant. We also find from Eq. (2–9) that $\mathbf{N}_u = -\kappa \mathbf{x}_u$, $\mathbf{N}_v = -\kappa \mathbf{x}_v$; here we use the fact that $e/E = f/F = g/G = \kappa$. If we now associate with a point $(\mathbf{x})$ on the surface a point

$$\mathbf{y} = \mathbf{x} + R\mathbf{N}, \qquad R = \kappa^{-1}, \quad \text{constant},$$

then $\mathbf{y}_u = \mathbf{y}_v = 0$, and all vectors $\mathbf{x} + R\mathbf{N}$ going out from the points on the surface meet in the fixed point $(\mathbf{y})$. Hence we conclude that

$$(\mathbf{x} - \mathbf{y}) \cdot (\mathbf{x} - \mathbf{y}) = R^2,$$

which is the equation of a sphere. When $\kappa = 0$ we find that $\mathbf{N}$ is a constant vector, and $d\mathbf{x} \cdot \mathbf{N} = 0$ can be integrated into $\mathbf{x} \cdot \mathbf{N} = \text{constant}$, the equation of a plane (which can be considered as a sphere of infinite radius).

(3) This result is related to the following theorem due to D. Hilbert:

In a region R of a surface of constant positive Gaussian curvature without umbilics the principal curvatures take their extreme values at the boundary.

Since the surface is not a sphere $\kappa_1 \neq \kappa_2$, so that we take $\kappa_1 = K/\kappa_2$. Let κ_1 assume a maximum at a point P inside the region R. Then κ_2 assumes a minimum at P. Such a point is not an umbilic and is a regular point of the surface, so that the region near P is covered simply and without gaps by lines of curvature, which we can take as coordinate lines. Hence at P, $\partial\kappa_1/\partial v = 0$ and $\partial\kappa_2/\partial u = 0$, and hence, according to the Codazzi Eqs. (3–8), $E_v = G_u = 0$. Moreover $\partial^2\kappa_1/\partial v^2 \leqslant 0$ and $\partial^2\kappa_2/\partial u^2 \geqslant 0$; hence, differentiating Eqs. (3–8), we obtain, since $\kappa_1 > \kappa_2$, that $E_{vv} \geqslant 0$, $G_{uu} \geqslant 0$. But K, at this point P, can be written, according to Eq. (3–6):

$$K = -\frac{1}{2EG}(E_{vv} + G_{uu}),$$

which has a right-hand member $\leqslant 0$ and a left-hand member > 0 ($E > 0$, $G > 0$). This is impossible, so that nowhere in the interior of the region can we have $\partial\kappa_1/\partial v = \partial\kappa_2/\partial u = 0$. The normal curvatures must monotonically increase or decrease along the lines of curvature.

The proof requires the existence of a sufficient number of continuous derivatives, since the only criterion used for an extremum of κ_1 and κ_2 is the vanishing of their first derivatives.

This result leads to the

Theorem of Liebmann. The only closed surface of constant positive curvature (without singularities) is the sphere.

Indeed, when a surface of constant curvature is not closed and is not a sphere, the maximum of all larger principal curvatures lies on the boundary. But a closed surface has no boundary, so that the larger principal curvatures must be equal at all points of the surface. This, however, means that both κ_1 and κ_2 must be constant on the surface, which can only happen (Eq. (3–8)) if all points are umbilics. The only possibility is therefore the sphere.

This proof again requires that at all points on the surface a sufficient number of continuous derivatives of $\mathbf{x}$ with respect to u and v exist. Indeed, when we take two congruent spherical caps and place them together in a symmetrical position along their boundary circle, we obtain a closed surface of constant positive curvature which is not a sphere, but which has a line of singular points.

This is our first example of *surface theory in the large.* We can express the theorem of Liebmann by saying that *a sphere cannot be bent.* A sphere with a hole in it can be bent, and we can also "crack" a sphere.

This theorem was suggested by F. Minding, Crelle's *Journal für Mathem.* **18**, 1838, p. 368; the first proof dates from H. Liebmann, *Göttinger Nachrichten,* 1899, pp. 44–55. The present proof follows W. Blaschke, *Differentialgeometrie*

I, sect. 91. Hilbert's proofs are in *Trans. Amer. Math. Soc.* **2**, 1901, pp. 97–99, also *Grundlagen der Geometrie, Anhang V*.

This theorem of Liebmann is a special case of the general theorem that a closed and convex surface cannot be bent, which theorem seems to go back to Lagrange (1812). See A. Cauchy, *Comptes Rendus* **21**, 1845, p. 564. A proof of this theorem was also given by Liebmann. It can also be proved that convex polyhedrons cannot be bent.

3–6 The fundamental theorem of surface theory. We have seen that the coefficients of the two quadratic forms

$$\mathrm{I} \equiv E\,du^2 + 2F\,du\,dv + G\,dv^2, \qquad EG - F^2 \neq 0, \qquad (6\text{–}1)$$
$$\mathrm{II} \equiv e\,du^2 + 2f\,du\,dv + g\,dv^2, \qquad\qquad\qquad (6\text{–}2)$$

satisfy the Gauss-Codazzi equations. We owe to O. Bonnet the proof of the converse theorem, called the

FUNDAMENTAL THEOREM. *If E, F, G and e, f, g are given as functions of u and v, sufficiently differentiable, which satisfy the Gauss-Codazzi equations (3–4) and (3–6), while $EG - F^2 \neq 0$, then there exists a surface which admits as its first and second fundamental forms $\mathrm{I} \equiv E\,du^2 + 2F\,du\,dv + G\,dv^2$ and $\mathrm{II} \equiv e\,du^2 + 2f\,du\,dv + g\,dv^2$ respectively. This surface is uniquely determined except for its position in space.* For real surfaces with real curvilinear coordinates (u, v) we need $EG - F^2 > 0$, $E > 0, G > 0$.

The demonstration consists in showing that the Gauss-Weingarten equations (equivalent to fifteen scalar equations):

$$\begin{aligned}
\mathbf{x}_{uu} &= \Gamma_{11}^1\mathbf{x}_u + \Gamma_{11}^2\mathbf{x}_v + e\mathbf{N}, & \text{(a)} \\
\mathbf{x}_{uv} &= \Gamma_{12}^1\mathbf{x}_u + \Gamma_{12}^2\mathbf{x}_v + f\mathbf{N}, & \text{(b)} \\
\mathbf{x}_{vv} &= \Gamma_{22}^1\mathbf{x}_u + \Gamma_{22}^2\mathbf{x}_v + g\mathbf{N}, & \text{(c)} \\
(EG - F^2)\mathbf{N}_u &= (fF - eG)\mathbf{x}_u + (eF - fE)\mathbf{x}_v, & \text{(d)} \\
(EG - F^2)\mathbf{N}_v &= (gF - fG)\mathbf{x}_u + (fF - gE)\mathbf{x}_v, & \text{(e)}
\end{aligned} \qquad (6\text{–}3)$$

under the additional conditions (validity at one point suffices)

$$\mathbf{N} \cdot \mathbf{N} = 1, \quad \mathbf{x}_u \cdot \mathbf{N} = 0, \quad \mathbf{x}_v \cdot \mathbf{N} = 0, \quad \mathbf{x}_u \cdot \mathbf{x}_u = E, \quad \mathbf{x}_u \cdot \mathbf{x}_v = F,$$
$$\mathbf{x}_v \cdot \mathbf{x}_v = G, \quad \mathbf{x}_{uu} \cdot \mathbf{N} = e, \quad \mathbf{x}_{uv} \cdot \mathbf{N} = f, \quad \mathbf{x}_{vv} \cdot \mathbf{N} = g. \quad (6\text{–}4)$$

determine the twelve scalar functions of u and v given by $\mathbf{x}, \mathbf{x}_u, \mathbf{x}_v, \mathbf{N}$ with just such a number of integration constants that the surface $\mathbf{x} = \mathbf{x}(u, v)$ is determined but for its position in space. Instead of a general demonstration of this theorem, which requires some knowledge of the integration theory of mixed systems of partial differential equations, we shall illustrate

the nature of this demonstration by means of a simple example as follows:
Given the differential forms

$$I \equiv du^2 + \cos^2 u \, dv^2,$$
$$II \equiv du^2 + \cos^2 u \, dv^2,$$

find the surface of which I *and* II *are the first and second fundamental forms.*

Since $E = 1$, $F = 0$, $G = \cos^2 u$; $e = 1$, $f = 0$, $g = \cos^2 u$, we find for the Christoffel symbols

$$\Gamma^1_{11} = \Gamma^2_{22} = \Gamma^1_{12} = \Gamma^2_{22} = 0, \qquad \Gamma^2_{12} = -\tan u, \qquad \Gamma^1_{22} = \sin u \cos u,$$

which satisfy the Gauss-Codazzi equations (3-3), (3-4), as direct substitution shows.

The Gauss-Weingarten equations (6-3) are in this case

$$\begin{align}
\mathbf{x}_{uu} &= \mathbf{N}, & \text{(a)} \\
\mathbf{x}_{uv} &= -\tan u \, \mathbf{x}_v, & \text{(b)} \\
\mathbf{x}_{vv} &= \sin u \cos u \, \mathbf{x}_u + \cos^2 u \, \mathbf{N}, & \text{(c)} \qquad \text{(6-5)} \\
\mathbf{N}_u &= -\mathbf{x}_u, & \text{(d)} \\
\mathbf{N}_v &= -\mathbf{x}_v. & \text{(e)}
\end{align}$$

We have to solve these 15 equations for the x_i. Eqs. (6-5a) and (6-5d) give by elimination of $\mathbf{N}$:

$$\mathbf{x}_{uuu} + \mathbf{x}_u = 0,$$

or

$$\mathbf{x} = \mathbf{a}(v) \cos u + \mathbf{b}(v) \sin u + \mathbf{c}(v).$$

Then Eq. (6-5b) gives

$$\mathbf{b}' \cos u = -\mathbf{b}' \sin^2 u \sec u - \mathbf{c}' \tan u, \quad (\mathbf{b}' = d\mathbf{b}/dv, \mathbf{c}' = d\mathbf{c}/dv)$$

or

$$\mathbf{b}' = \mathbf{c}' = 0.$$

Hence $\mathbf{b}$ and $\mathbf{c}$ are constant vectors, and

$$\mathbf{x} = \mathbf{a}(v) \cos u + \mathbf{b} \sin u + \mathbf{c}.$$

From this equation and from Eqs. (6-5c), (6-5b), and (6-5e), we obtain:

$$\mathbf{x}_{vvv} = \mathbf{a}''' \cos u = \sin u \cos u \, \mathbf{x}_{uv} + \cos^2 u \, \mathbf{N}_v = -\mathbf{a}' \cos u,$$

or

$$\mathbf{a}''' + \mathbf{a}' = 0,$$
$$\mathbf{a} = \mathbf{p} \cos v + \mathbf{q} \sin v + \mathbf{r}, \qquad \mathbf{p}, \mathbf{q}, \mathbf{r} \text{ constant vectors.}$$

The solution of Eq. (6-3) is therefore [$\mathbf{r} = 0$, according to Eq. 6-5(c)]

$$\mathbf{x} = \mathbf{p} \cos v \cos u + \mathbf{q} \sin v \cos u + \mathbf{b} \sin u + \mathbf{c}.$$

We must now select $\mathbf{p}$, $\mathbf{q}$, $\mathbf{b}$, $\mathbf{c}$ so as to satisfy the additional conditions (6–4). Since

$$\mathbf{x}_u = -\mathbf{p}\cos v \sin u - \mathbf{q} \sin v \sin u + \mathbf{b} \cos u,$$
$$\mathbf{x}_v = -\mathbf{p}\sin v \cos u + \mathbf{q} \cos v \cos u,$$

we find that for all u and v:

$$\mathbf{x}_v \cdot \mathbf{x}_v = \cos^2 u = (\mathbf{p}\cdot\mathbf{p})\sin^2 v \cos^2 u - 2(\mathbf{p}\cdot\mathbf{q})\sin v \cos v \cos^2 u$$
$$+ (\mathbf{q}\cdot\mathbf{q})\cos^2 v \cos^2 u,$$

or

$$\mathbf{p}\cdot\mathbf{p} = 1, \qquad \mathbf{p}\cdot\mathbf{q} = 0, \qquad \mathbf{q}\cdot\mathbf{q} = 1.$$

Similarly we derive from $\mathbf{x}_u \cdot \mathbf{x}_v = 0$:

$$\mathbf{q}\cdot\mathbf{b} = \mathbf{p}\cdot\mathbf{b} = 0.$$

Then we obtain from $\mathbf{x}_u \cdot \mathbf{x}_u = 1$:

$$\mathbf{b}\cdot\mathbf{b} = 1.$$

This shows that $\mathbf{p}$, $\mathbf{q}$, $\mathbf{b}$ form a system of mutually orthogonal unit vectors. The other equations (6–4) are now automatically satisfied. Denoting the vectors $\mathbf{p}$, $\mathbf{q}$, $\mathbf{b}$ by $\mathbf{e}_1$, $\mathbf{e}_2$, $\mathbf{e}_3$ respectively, we obtain as the general solution of our set of equations:

$$\mathbf{x} = \mathbf{e}_1 \cos v \cos u + \mathbf{e}_2 \sin v \cos u + \mathbf{e}_3 \sin u + \mathbf{c},$$

which is the equation of the unit sphere. By the choice of $\mathbf{e}_1$, $\mathbf{e}_2$, $\mathbf{e}_3$, $\mathbf{c}$ we can place this sphere in any position of space, selecting any orthogonal system of meridians and parallels for our u- and v-coordinates.

The first to give a proof of the fundamental theorem was O. Bonnet, *Journal Ecole Polytechnique*, cah. **42** (1867); there are also proofs in Bianchi's *Lezioni*, section 68, and in Eisenhart's *Differential geometry*, pp. 158–159. The reader may compare this fundamental theorem with that for space curves, in which we consider a system of ordinary differential equations, which does not need conditions of the Gauss-Codazzi type for complete integrability. In the case of curves, we have given an actual method to carry out the integration (Section 1–10), which leads to a Riccati equation. The analogous problem in the case of surfaces can again be reduced to a Riccati equation. A complete discussion of this case is in G. Scheffers, *Anwendung II*, pp. 393–414. The general theory of mixed systems of partial differential equations on which the proof of the fundamental theorem is based can be studied in T. Levi Civita, *The absolute differential calculus*, Chap. 2, Sec. 8.

CHAPTER 4

GEOMETRY ON A SURFACE

4-1 Geodesic (tangential) curvature. We are now able to continue our study of the intrinsic properties of the surface. As in Section 2-5 we start with the curvature vector of a curve C on the surface. This curvature vector

$$dt/ds = \kappa \mathbf{n} = \mathbf{k}$$

of the curve at a point P with the tangent direction $\mathbf{t}$ lies in the plane through P perpendicular to $\mathbf{t}$. This plane also contains the surface normal, in which the unit vector $\mathbf{N}$ is laid (Fig. 4-1). The projection of the curvature vector $\mathbf{k}$ on the surface normal is, according to Meusnier's theorem (Section 2-5), the curvature vector of the normal section in direction $\mathbf{t}$; we have called this vector $\mathbf{k}_n$. We shall now study the projection of $\mathbf{k}$ on the tangent plane, which is called the *vector of tangential curvature* and which we have denoted by $\mathbf{k}_g$. Hence we have the relation, already expressed in Eq. (5-1), Chapter 2:

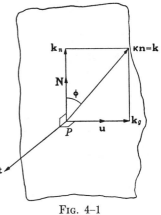

Fig. 4-1

$$\boxed{\mathbf{k} = \mathbf{k}_n + \mathbf{k}_g}, \tag{1-1}$$

in words:

The curvature vector is the sum of the normal and the tangential curvature vectors.

The tangential curvature vector is also called the *geodesic curvature vector*. To find an expression for $\mathbf{k}_g$ we introduce a unit vector $\mathbf{u}$ perpendicular to $\mathbf{t}$ in the tangent plane in such a way that the sense $\mathbf{t} \rightarrow \mathbf{u}$ is the same as that of $\mathbf{x}_u \rightarrow \mathbf{x}_v$. We then introduce the *tangential curvature* or *geodesic curvature* κ_g by means of the equation

$$\mathbf{k}_g = \kappa_g \mathbf{u}. \tag{1-2}$$

Where Meusnier's theorem gives us $|\kappa \cos \varphi|$ for the magnitude κ_n of the normal curvature vector $\mathbf{k}_n$, we find $|\kappa_g| = |\kappa \sin \varphi|$ as the magnitude of the geodesic curvature vector. We shall now show that this vector, contrary to $\mathbf{k}_n$, depends only on E, F, G and their derivatives (as well as on the function $\psi(u, v) = 0$ defining the curve C), and is therefore a bending invariant.

From Eq. (1–1) follows, since $\mathbf{u} \cdot \mathbf{k}_n = \mathbf{u} \cdot \kappa_n\mathbf{N} = 0$:

$$\kappa_g = \mathbf{u} \cdot (d\mathbf{t}/ds) = \mathbf{u} \cdot \mathbf{t}'.$$

Hence, since $\mathbf{t} \rightarrow \mathbf{u} \rightarrow \mathbf{N}$ has the positive (right-handed) sense:

$$\kappa_g = (\mathbf{N} \times \mathbf{t}) \cdot \mathbf{t}' = (\mathbf{t}\mathbf{t}'\mathbf{N}). \tag{1–3}$$

The unit vector $\mathbf{t}$ satisfies the equation

$$\mathbf{t} = \mathbf{x}_u u' + \mathbf{x}_v v', \qquad u' = du/ds, \qquad v' = dv/ds;$$

hence

$$\mathbf{t}' = \mathbf{x}_{uu}(u')^2 + 2\mathbf{x}_{uv}u'v' + \mathbf{x}_{vv}(v')^2 + \mathbf{x}_u u'' + \mathbf{x}_v v'', \tag{1–4}$$

so that

$$\begin{aligned}(\mathbf{t}\mathbf{t}'\mathbf{N}) = [&(\mathbf{x}_u \times \mathbf{x}_{uu})(u')^3 + (2\mathbf{x}_u \times \mathbf{x}_{uv} + \mathbf{x}_v \times \mathbf{x}_{uu})u'^2v' \\ &+ (\mathbf{x}_u \times \mathbf{x}_{vv} + 2\mathbf{x}_v \times \mathbf{x}_{uv})u'(v')^2 + (\mathbf{x}_v \times \mathbf{x}_{vv})(v')^3] \cdot \mathbf{N} \\ &+ (\mathbf{x}_u \times \mathbf{x}_v) \cdot \mathbf{N}(u'v'' - u''v').\end{aligned} \tag{1–5}$$

The coefficients of $(u')^3$, $(u')^2v'$, etc. are all functions of E, F, G and their first derivatives. Indeed:

$$\begin{aligned}(\mathbf{x}_u \times \mathbf{x}_{uu}) \cdot \mathbf{N} &= \frac{(\mathbf{x}_u \times \mathbf{x}_{uu}) \cdot (\mathbf{x}_u \times \mathbf{x}_v)}{\sqrt{EG - F^2}} \\ &= \frac{(\mathbf{x}_u \cdot \mathbf{x}_u)(\mathbf{x}_{uu} \cdot \mathbf{x}_v) - (\mathbf{x}_u \cdot \mathbf{x}_{uu})(\mathbf{x}_u \cdot \mathbf{x}_v)}{\sqrt{EG - F^2}} \\ &= \frac{E[11, 2] - F[11, 1]}{\sqrt{EG - F^2}} \\ &= \Gamma_{11}^2\sqrt{EG - F^2}.\end{aligned}$$

Similarly:

$$(\mathbf{x}_u \times \mathbf{x}_{uv}) \cdot \mathbf{N} = \Gamma_{12}^2\sqrt{EG - F^2}, \text{ etc.}$$
$$(\mathbf{x}_u \times \mathbf{x}_v) \cdot \mathbf{N} = \sqrt{EG - F^2}.$$

Therefore, according to Eqs. (1–3) and (1–5):

$$\begin{aligned}\kappa_g = [\Gamma_{11}^2(u')^3 &+ (2\Gamma_{12}^2 - \Gamma_{11}^1)(u')^2v' + (\Gamma_{22}^2 - 2\Gamma_{12}^1)u'(v')^2 - \Gamma_{22}^1(v')^3 \\ &+ u'v'' - u''v']\sqrt{EG - F^2}, \tag{1–6}\end{aligned}$$

where the square root is taken positive, and this equation shows that κ_g depends on only E, F, G, their first derivatives, and on u', u'', v', v'':

The tangential curvature of a curve is a bending invariant.

Since $\mathbf{k}_g$ lies in the tangent plane and is perpendicular to $\mathbf{t}$, it is also a bending invariant.

When we apply Eq. (1–6) to a plane curve, taking $u = x$, $v = y$, $E = 1$, $F = 0$, $G = 1$, then we obtain

$$\kappa_g = u'v'' - u''v', \tag{1–7}$$

which means that $\kappa_g = d\varphi/ds$. Our choice of the sign of κ_g therefore makes κ_g the ordinary curvature of a plane curve (p. 15). The tangential curvature vector itself is unaffected by the choice of sign of κ_g. It depends only on the geometrical structure of the curve. Formula (1–1) can now be written (comp. Chapter 2, Eq. 5–2):

$$\mathbf{k} = \kappa_n \mathbf{N} + \kappa_g \mathbf{u}. \tag{1–8}$$

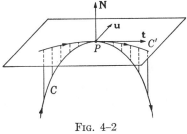

When we construct the cylinder which projects the curve C on the tangent plane at P, then the generating lines of this cylinder have the direction of $\mathbf{N}$, and the normal to the cylinder at P has the direction of $\mathbf{u}$ (Fig. 4–2). The geodesic curvature

Fig. 4–2

vector is therefore the normal curvature vector of C as curve on the projecting cylinder, hence the curvature vector of the normal section of the cylinder in the direction of $\mathbf{t}$. This normal section is the projection C' of C on the tangent plane. This means:

The tangential curvature vector of a curve at P is the (ordinary) curvature vector of the projection on the tangent plane of the surface at P.

EXAMPLE: The great circles on a sphere have geodesic curvature zero, since they project on the tangent plane at each of their points as straight lines.

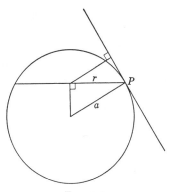

The small circle of radius r on a sphere with radius a projects on the tangent plane at one of its points as an ellipse of semi-axes $A = r$ and $B = r\sqrt{a^2-r^2}/a$. The geodesic curvature of such a small circle is therefore constant and equal to the curvature of this ellipse at the vertex of the minor axis, which is equal to $B/A^2 = \sqrt{a^2 - r^2}/ar$ (Fig. 4–3). This is equal to the curvature of the circle obtained by developing the small circle on the plane as a curve on the cone which is tangent to the sphere along the small circle (see Exercise 2, Section 4–8).

Fig. 4–3

The geodesic curvature was already known to Gauss. The first publication on this subject was by F. Minding, Crelle's *Journal für Mathem.* **5**, 1830, p. 297, who in the next issue (Vol. 6, 1830, p. 159) showed its character as a bending invariant. The name geodesic curvature is due to O. Bonnet, *Journal Ecole Polytechn.* **19**, 1848, p. 43. Ferdinand Minding (1806–1885), whose name we shall repeatedly meet, was a professor at the German university of Dorpat (now Tartu, Estonia).

We present here for further use the expressions for the geodesic curvature of the parametric lines:

$$(\kappa_g)_{v=\text{const}} = \Gamma_{11}^2 (u')^3 \sqrt{EG - F^2} = \Gamma_{11}^2 \frac{\sqrt{EG - F^2}}{E\sqrt{E}} \left(\text{since } u' = \frac{1}{\sqrt{E}} \right),$$

$$(\kappa_g)_{u=\text{const}} = -\Gamma_{22}^1 (v')^3 \sqrt{EG - F^2} = -\Gamma_{22}^1 \frac{\sqrt{EG - F^2}}{G\sqrt{G}}. \tag{1-9}$$

In particular, when the parametric lines are orthogonal, we have $\Gamma_{11}^2 = -\frac{1}{2} E_v/G$, $\Gamma_{22}^1 = -\frac{1}{2} G_u/E$; and s_1 and s_2 are taken along the respective parametric lines,

$$(\kappa_g)_1 = (\kappa_g)_{v=\text{const}} = -\frac{1}{2} \frac{E_v}{E\sqrt{G}} = -\frac{1}{\sqrt{G}} \frac{\partial \ln\sqrt{E}}{\partial v} = -\frac{d}{ds_2} \ln\sqrt{E},$$

$$(\kappa_g)_2 = (\kappa_g)_{u=\text{const}} = +\frac{1}{2} \frac{G_u}{G\sqrt{E}} = +\frac{1}{\sqrt{E}} \frac{\partial \ln\sqrt{G}}{\partial u} = +\frac{d}{ds_1} \ln\sqrt{G}. \tag{1-10}$$

For polar coordinates in the plane, where $E = 1$, $G = r^2 = u^2$, we find for the circles $r = \text{const}$, $\kappa_g = +r^{-1}$, which illustrates again the choice of the sign.

Let us now introduce the unit vectors $\mathbf{i}_1$ and $\mathbf{i}_2$ along the orthogonal parametric lines. Then, according to Eq. (1–8):

$$(\kappa_g)_1 = \mathbf{i}_2 \cdot \frac{d\mathbf{i}_1}{ds_1}, \qquad (\kappa_g)_2 = -\mathbf{i}_1 \cdot \frac{d\mathbf{i}_2}{ds_2}, \tag{1-11}$$

where s_1 and s_2 are taken along the respective parametric lines. We now consider a curve C through P, $\mathbf{x} = \mathbf{x}(s)$, making an angle θ with the curve $v = \text{constant}$. Then its unit tangent vector $\mathbf{t}$ satisfies the equation:

$$\mathbf{t} = \mathbf{i}_1 \cos\theta + \mathbf{i}_2 \sin\theta.$$

Differentiating $\mathbf{t}$ along C and using the formula

$$\frac{d\mathbf{i}_\alpha}{ds} = \frac{\partial \mathbf{i}_\alpha}{\partial u} \frac{du}{ds} + \frac{\partial \mathbf{i}_\alpha}{\partial v} \frac{dv}{ds} = \frac{d\mathbf{i}_\alpha}{ds_1} \cos\theta + \frac{d\mathbf{i}_\alpha}{ds_2} \sin\theta \; (\alpha = 1, 2),$$

we obtain for the curvature vector of C:

$$\frac{d\mathbf{t}}{ds} = \frac{d\mathbf{i}_1}{ds_1} \cos^2 \theta + \frac{d\mathbf{i}_1}{ds_2} \cos \theta \sin \theta + \frac{d\mathbf{i}_2}{ds_1} \cos \theta \sin \theta + \frac{d\mathbf{i}_2}{ds_2} \sin^2 \theta + \mathbf{u} \frac{d\theta}{ds},$$

where, according to the definition (comp. Eq. (1–2)),

$$\mathbf{u} = -\mathbf{i}_1 \sin \theta + \mathbf{i}_2 \cos \theta. \tag{1–12}$$

Hence, by virtue of Eqs. (1–2) and (1–12) we obtain for the geodesic curvature κ_g of C:

$$\kappa_g = \frac{d\theta}{ds} - \mathbf{i}_1 \cdot \frac{d\mathbf{i}_2}{ds_1} \cos \theta \sin^2 \theta - \mathbf{i}_1 \cdot \frac{d\mathbf{i}_2}{ds_2} \sin^3 \theta + \mathbf{i}_2 \cdot \frac{d\mathbf{i}_1}{ds_1} \cos^3 \theta$$
$$+ \mathbf{i}_2 \cdot \frac{d\mathbf{i}_1}{ds_2} \cos^2 \theta \sin \theta,$$

or, because of Eq. (1–11):

$$\kappa_g = \frac{d\theta}{ds} + (\kappa_g)_1 \cos \theta \sin^2 \theta + (\kappa_g)_2 \sin^3 \theta + (\kappa_g)_1 \cos^3 \theta + (\kappa_g)_2 \cos^2 \theta \sin \theta,$$

or finally

$$\boxed{\kappa_g = \frac{d\theta}{ds} + (\kappa_g)_1 \cos \theta + (\kappa_g)_2 \sin \theta.} \tag{1–13}$$

This formula is known as *Liouville's formula* for the geodesic curvature. It can be found in a note to Liouville's edition of Monge's *Applications* (1850).

4–2 Geodesics. Geodesic lines are sometimes defined as lines of shortest distance between points on a surface. This is not always a satisfactory definition, and we shall therefore define geodesics in a different way, postponing the discussion of the minimum property. *Our definition of geodesics is that they are curves of zero geodesic curvature.*

This means that straight lines on the surface are geodesics, since in this case the curvature vector $\mathbf{k}$ vanishes. For all curved geodesics our definition means that the curvature vector at each point coincides with the normal curvature vector. In other words, *the osculating planes of a curved geodesic contain the surface normal.* We can therefore state the following property:

All straight lines on a surface are geodesics. Along all curved geodesics the principal normal coincides with the surface normal.

Along (curved) asymptotic lines osculating planes and tangent planes coincide, along (curved) geodesics they are normal. Through a point of a nondevelopable surface pass two asymptotic lines (real or imaginary). Through a point of a surface passes a geodesic in every tangent direction.

This last property becomes plausible when we realize that we can find at a point on the surface in every direction a curve of which the osculating plane passes through the surface normal. It can be shown analytically by considering the equation of the geodesic lines. We obtain this equation by simply writing $\kappa_g = 0$ in Eq. (1–6) and obtain

$$u'v'' - u''v' = -\Gamma_{11}^2 (u')^3 - (2\Gamma_{12}^2 - \Gamma_{11}^1)(u')^2 v' + (2\Gamma_{12}^1 - \Gamma_{22}^2)u'(v')^2 + \Gamma_{22}^1 (v')^3. \quad (2–1)$$

This equation is derived from Eq. (1–3), so that the accents indicate differentiation with respect to the arc length. However, since

$$\frac{du}{ds}\frac{d^2v}{ds^2} - \frac{dv}{ds}\frac{d^2u}{ds^2} = \left[\frac{du}{dt}\frac{d^2v}{dt^2} - \frac{dv}{dt}\frac{d^2u}{dt^2}\right]\left(\frac{dt}{ds}\right)^3,$$

this equation (2–1) is still correct when the accents denote differentiation with respect to any parameter.

It is often convenient to express the geodesics in another way. This can be accomplished by noting that along these lines **N** has the direction of $\pm \mathbf{n}$. Hence,

$$\mathbf{n} \cdot \mathbf{x}_u = 0, \qquad \mathbf{n} \cdot \mathbf{x}_v = 0,$$

or (differentiation is again with respect to s):

$$\mathbf{t}' \cdot \mathbf{x}_u = 0, \qquad \mathbf{t}' \cdot \mathbf{x}_v = 0,$$

which equations (compare with Eq. (1–4)) are equivalent to

$$(\mathbf{x}_{uu} \cdot \mathbf{x}_u)(u')^2 + 2(\mathbf{x}_{uv} \cdot \mathbf{x}_u)u'v' + (\mathbf{x}_{vv} \cdot \mathbf{x}_u)(v')^2 + Eu'' + Fv'' = 0,$$
$$(\mathbf{x}_{uu} \cdot \mathbf{x}_v)(u')^2 + 2(\mathbf{x}_{uv} \cdot \mathbf{x}_v)u'v' + (\mathbf{x}_{vv} \cdot \mathbf{x}_v)(v')^2 + Fu'' + Gv'' = 0.$$

Eliminating v'' from these equations and also u'', substituting $\mathbf{x}_{uu} \cdot \mathbf{x}_u = [11, 1]$, and then introducing the Christoffel symbols from Section 3–2, *we obtain the equation of the geodesics in the form:*

$$\boxed{\begin{aligned}\frac{d^2u}{ds^2} + \Gamma_{11}^1\left(\frac{du}{ds}\right)^2 + 2\Gamma_{12}^1\frac{du}{ds}\frac{dv}{ds} + \Gamma_{22}^1\left(\frac{dv}{ds}\right)^2 = 0,\\[4pt] \frac{d^2v}{ds^2} + \Gamma_{11}^2\left(\frac{du}{ds}\right)^2 + 2\Gamma_{12}^2\frac{du}{ds}\frac{dv}{ds} + \Gamma_{22}^2\left(\frac{dv}{ds}\right)^2 = 0.\end{aligned}} \quad (2–2)$$

The history of geodesic lines begins with John Bernoulli's solution of the problem of the shortest distance between two points on a convex surface (1697–1698). His answer was that the osculating plane ("the plane passing through three points 'quolibet proxima'") must always be perpendicular to the tangent plane. For the further history see P. Stäckel, *Bemerkungen zur Geschichte der geodätischen Linien*, Berichte sächs. Akad. Wiss., Leipzig, **45**, 1893, pp. 444–467.

The name "geodesic line" in its present meaning is, according to Stäckel, due to J. Liouville, *Journal de mathém.* **9**, 1844, p. 401; the equation of the geodesics was first obtained by Euler in his article *De linea brevissima in superficie quacumque duo quaelibet puncta jungente, Comment. Acad. Petropol.* **3** (ad annum 1728), 1732; Euler's equation refers to a surface given by $F(x, y, z) = 0$ (see Exercise 22, this section).

It seems at first strange that Eqs. (2–2) give two conditions, whereas $\kappa_g = 0$ is only one condition. However, Eqs. (2–2) also express only one condition, since they are related by the relation $ds^2 = E\, du^2 + 2F\, du\, dv + G\, dv^2$. When we eliminate ds from (2–2) we obtain again one equation for the geodesics, now in the form

$$\frac{d^2v}{du^2} = \Gamma_{22}^1 \left(\frac{dv}{du}\right)^3 + (2\Gamma_{12}^1 - \Gamma_{22}^2)\left(\frac{dv}{du}\right)^2 + (\Gamma_{11}^1 - 2\Gamma_{12}^2)\frac{dv}{du} - \Gamma_{11}^2. \quad (2\text{–}3a)$$

This equation can immediately be found from Eq. (2–1), taking u as parameter. It is of the form:

$$v'' = A(v')^3 + B(v')^2 + Cv' + D, \quad\quad (2\text{–}3b)$$

where A, B, C, D are functions of u and v.

From Eqs. (2–3a, b) we see that when at a point $P(u, v)$ a direction dv/du is given, d^2v/du^2 is determined, that is, the way in which the curve of given direction is continued. More precisely, using the existence theorem for the solutions of ordinary differential equations of the second order:

Through every point of the surface passes a geodesic in every direction.

A geodesic is uniquely determined by an initial point and tangent at that point.

The existence theorem states that a differential equation

$$d^2v/du^2 = f(u, v, dv/du),$$

where f is a single-valued and continuous function of its independent variables inside a certain given interval (with a Lipschitz condition satisfied in our case), has inside this interval a unique continuous solution for which dv/du takes a given value $(dv/du)_0$ at a point $P(u_0, v_0)$.[*]

EXAMPLES. 1. *Plane.* The straight lines in the plane satisfy Eqs. (2–2) or (2–3); they are the geodesics in the plane. Through every point passes a line in every direction.

---

[*] Compare P. Franklin, *Treatise on advanced calculus*, Wiley, N. Y., 1940, pp. 518–522.

2. *Sphere.* The osculating planes of the geodesics all pass through the center, and the geodesics are therefore all curves in planes through the center (see Exercise 3, Section 1–6). *The great circles of a sphere are its geodesics.*

3. *Surface of revolution.* Here (Section 2–2) we have,

$$E = 1 + f'^2, \qquad F = 0, \qquad G = u^2, \qquad f = f(u),$$

hence

$$\Gamma_{11}^1 = \frac{f'f''}{1 + (f')^2}, \qquad \Gamma_{12}^2 = \frac{1}{u}, \qquad \Gamma_{22}^1 = -\frac{u}{1 + f'^2}, \qquad \Gamma_{11}^2 = \Gamma_{12}^1 = \Gamma_{22}^2 = 0.$$

The equations of the geodesics (2–2) are here:

$$\frac{d^2u}{ds^2} + \frac{f'f''}{1 + (f')^2}\left(\frac{du}{ds}\right)^2 - \frac{u}{1 + f'^2}\left(\frac{dv}{ds}\right)^2 = 0, \qquad \frac{d^2v}{ds^2} + \frac{2}{u}\frac{du}{ds}\frac{dv}{ds} = 0.$$

One of these equations suffices; we take the second:

$$\frac{d^2v/ds^2}{dv/ds} + 2\,\frac{du/ds}{u} = 0, \qquad u^2\left(\frac{dv}{ds}\right) = c, \qquad c = \text{constant.}$$

This value substituted into the equation for ds^2 gives

$$u^4\,dv^2 = c^2(1 + f'^2)\,du^2 + c^2u^2\,dv^2,$$
$$v = \pm c \int \frac{\sqrt{1 + (f')^2}}{u\sqrt{u^2 - c^2}}\,du. \tag{2–4}$$

The geodesics of a surface of revolution can be found by quadratures.

The value $c = 0$ gives $v = $ constant; the meridians. The parallels $u = $ constant are geodesics when $f' = \infty$, which means that along such parallels the tangent planes envelop a cylinder with generating lines parallel to the axis.

EXERCISES

1. Find the geodesics of the plane by integrating Eq. (2–2) in polar coordinates.

2. Find the geodesics of the sphere by integrating Eq. (2–2) in spherical coordinates.

3. Find the geodesics on a right circular cone.

4. Show that the geodesics of the right helicoid can be found by means of elliptic integrals.

5. Find the equation of the geodesics when $z = f(x, y)$.

6. The geodesics on cylinders are helices.

7. *Theorem of Clairaut.* When a geodesic line on a surface of revolution makes an angle α with the meridian, then along the geodesic $u \sin \alpha = $ constant, where u is the radius of the parallel. (A. C. Clairaut, *Mém. Acad. Paris pour 1733*, 1735, p. 86.)

8. Show that the evolutes of a curve are geodesics on the polar developable.

9. When a curve is (a) asymptotic, (b) geodesic, (c) straight, show that (a) and (b) involve (c). Do (a) and (c) involve (b)? Do (b) and (c) involve (a)?

10. When a curve is (a) geodesic, (b) a line of curvature, (c) plane, show that (a) and (b) involve (c). Do (a) and (c) involve (b)? Do (b) and (c) involve (a)?

11. Prove that for a curve $\mathbf{x} = \mathbf{x}(s)$ with tangent direction given by the unit vector $\mathbf{t}$, $D\kappa_g = \dfrac{\partial}{\partial u}(\mathbf{t} \cdot \mathbf{x}_v) - \dfrac{\partial}{\partial v}(\mathbf{t} \cdot \mathbf{x}_u)$. $(D = \sqrt{EG - F^2}.)$

12. Use the formula of Exercise 11 to prove Liouville's formula.

13. Show that when the parametric lines are orthogonal

$$K = \frac{1}{\sqrt{EG}}\left[\frac{\partial}{\partial v}(\kappa_1\sqrt{E}) - \frac{\partial}{\partial u}(\kappa_2\sqrt{G})\right],$$

$$K = \frac{d\kappa_1}{ds_1} - \frac{d\kappa_2}{ds_2} - \kappa_1^2 - \kappa_2^2.$$

(J. Liouville, *Journal de mathém.* **16**, 1851, p. 103.)

14. *Bonnet's formula.* When a curve on a surface is given by $\varphi(u, v) = $ constant, show that

$$\kappa_g = \frac{1}{\sqrt{EG - F^2}}\left[\frac{\partial}{\partial u}\frac{F\varphi_v - G\varphi_u}{\sqrt{E\varphi_v^2 - 2F\varphi_u\varphi_v + G\varphi_u^2}} + \frac{\partial}{\partial v}\frac{F\varphi_u - E\varphi_v}{\sqrt{E\varphi_v^2 - 2F\varphi_u\varphi_v + G\varphi_u^2}}\right].$$

(O. Bonnet, *Paris Comptes Rendus* **42**, 1856, p. 1137.)

15. The curvature of a line of curvature is, but for the sign, equal to its normal curvature times the curvature of its spherical representation.

16. Show that the geodesics on a torus $x = r\cos\varphi$, $y = r\sin\varphi$, $z = a\sin\theta$ are given by (c is an arbitrary constant, $r = p + a\cos\theta$):

$$d\varphi = \frac{ca\,dr}{r\sqrt{r^2 - c^2}\sqrt{a^2 - (r - p)^2}}.$$

17. Show that the finite equation of the geodesics on a paraboloid of revolution can be written with the aid of elementary functions.

18. *Surfaces of Liouville.* There are surfaces of which the line element can be reduced to the form

$$ds^2 = (U + V)(du^2 + dv^2),$$

where U is a function of u alone and V of v alone. Show that the geodesics can be obtained by a quadrature. (J. Liouville, *Journal de Mathém.* **11**, 1846, p. 345.)

19. Show that surfaces of revolution are special cases of Liouville surfaces and that Clairaut's theorem (Exercise 7) may be extended to Liouville surfaces in the form

$$U\sin^2\omega - V\cos^2\omega = \text{constant},$$

where ω is the angle at which the geodesic cuts the curve $v = $ constant.

20. A surface for which the radii of principal curvature are constants is a sphere.

21. Prove that the compatibility conditions for the Weingarten equations (2–9), Chapter 3, lead to the Codazzi equations.

22. Show that the equation of a geodesic on a surface given by $F(x, y, z) = 0$ is

$$\begin{vmatrix} F_x & dx & d^2x \\ F_y & dy & d^2y \\ F_z & dz & d^2z \end{vmatrix} = 0, \qquad (F_x\,dx + F_y\,dy + F_z\,dz = 0).$$

4–3 Geodesic coordinates. Let us consider an arbitrary curve C_0 on a surface and the geodesics intersecting this curve at right angles (Fig. 4–4). This is quite general, since through a point on any curve C_0 passes one and only one geodesic in a direction perpendicular to the tangent; we consider only that part of the surface where the geodesics do not intersect. We take the geodesics as the parameter curves $v = $ constant, and the orthogonal trajectories of these geodesics as the curves $u = $ constant. The curve C_0 can be selected as $u = 0$. The ds^2 of the surface has no F:

$$ds^2 = E\,du^2 + G\,dv^2,$$

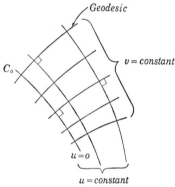

and ds^2 is further specialized by the condition that the geodesic curvature $(\kappa_g)_{v=\text{const}}$ vanishes. This, according to Eq. (1–10), means that $E_v = 0$:

$$ds^2 = E(u)\,du^2 + G(u, v)\,dv^2.$$

If we introduce a new parameter

$$u_1 = \int_0^u \sqrt{E}\,du,$$

the ds^2 assumes the form

Fig. 4–4

$$\boxed{ds^2 = du^2 + G(u, v)\,dv^2}, \qquad (3\text{–}1)$$

where we have replaced u_1 again by u; the parameter u now measures the arc length along the geodesic lines $v = $ constant, starting with $u = 0$ on C_0. Eq. (3–1) is called the *geodesic form* of the line element and (u, v) form a set of *geodesic coordinates;* or, for short, *geodesic set.* On every surface we can find an infinite number of geodesic sets, depending on an arbitrary curve C_0, along which the curves $v = $ constant can still be spaced in an arbitrary way.

Another way to introduce geodesic coordinates is to consider at an arbitrary point O on the surface (Fig. 4–5) the geodesics starting from this point in all directions ($v = $ constant), as well as their orthogonal directions ($u = $ constant). We find again for ds^2 the form (3–1), in which u is now the arc length along the geodesics, measured from O. In this case we speak of *geodesic polar coordinates*. They also form a geodesic set.

For the arc length of the curve $v = c_1$ between its intersections with the curves $u = u_1$, $u = u_2$ we find, according to Eq. (3–1):

$$s = \int_{u_1}^{u_2} du = u_2 - u_1, \tag{3–2}$$

which means that *the segments on all the geodesics $v = $ constant included between any two orthogonal trajectories are equal.* And conversely:

If geodesics be drawn orthogonal to a curve C, and segments of equal length be measured upon them from C, then the locus of their end points is an orthogonal trajectory of the geodesics.

This is the reason that the orthogonal trajectories of a system of ∞^1 geodesics are called *geodesic parallels*.

Geodesic coordinates were introduced and extensively used by Gauss in his *Disquisitiones* of 1827. Parallel curves in the plane were already studied by Leibniz in 1692, who called the involutes of a curve "parallel."

We have fixed only the parameter u (but for a constant), and can still introduce any function of v as a new parameter v_1 (which means that we can still space the geodesics in any desired way). In the first case, illustrated by Fig. 4–4, we take as the parameter v_1 the arc length along $C_0(u = 0)$, beginning with an arbitrary point on C_0 (for which we can take the intersection with $v = 0$):

$$v_1 = \int_0^v \sqrt{G(0, v)}\, dv.$$

Hence in the new parameter $G(0, v_1) = 1$. We still have C_0 itself to dispose of; we select it as a geodesic. Then $(\kappa_g)_{u=0} = 0$, or, according to Eq. (1–10):

$$\left[\frac{\partial}{\partial u} \sqrt{G(u, v)}\right]_{u=0} = 0.$$

In the second case, that of geodesic polar coordinates, we have as a first condition that at O: $G(0, v) = 0$. We now select as new parameter v_1 the angle which the geodesic $v = $ constant makes with the geodesic $v = 0$. This angle can be found most easily by realizing that the curves $u = $ constant

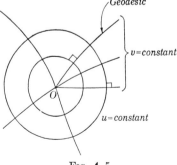

FIG. 4–5

for decreasing u (Fig. 4–5) approach ordinary circles, since for very small u the curve $u = u_0 = $ constant differs little from the circle of radius u_0 and center O in the tangent plane. This leads us to the equation (u being the arc length along the curves $v = $ constant):

$$\lim_{u \to 0} \frac{\int_0^{v_1} \sqrt{G(u, v_1)}\, dv_1}{u v_1} = 1,$$

or, applying l'Hôpital's rule,

$$\int_0^{v_1} \left[\frac{\partial}{\partial u} \sqrt{G(u, v_1)} \right]_{u=0} dv_1 = v_1 \quad \text{or} \quad \left[\frac{\partial}{\partial u} \sqrt{G(u, v_1)} \right]_{u=0} = 1.$$

We have thus found the following result:

We can always reduce the ds^2 of a surface to geodesic coordinates

$$ds^2 = du^2 + G(u, v)\, dv^2,$$

by a method (Fig. 4–4) which allows us to write

$$\sqrt{G(0, v)} = 1, \qquad \left[\frac{\partial}{\partial u} \sqrt{G(u, v)} \right]_{u=0} = 0 \qquad (3\text{–}3)$$

and also by a method (Fig. 4–5, geodesic polar coordinates) which allows us to write

$$\sqrt{G(0, v)} = 0, \qquad \left[\frac{\partial}{\partial u} \sqrt{G(u, v)} \right]_{u=0} = 1. \qquad (3\text{–}4)$$

Since for geodesic coordinates

$$\Gamma_{11}^1 = \Gamma_{11}^1 = \Gamma_{12}^1 = 0,$$

$$\Gamma_{12}^2 = \frac{1}{2} \frac{G_u}{G}, \qquad \Gamma_{22}^1 = -\frac{1}{2} G_u, \qquad \Gamma_{22}^2 = \frac{1}{2} \frac{G_v}{G},$$

we find for the geodesic curvature of the orthogonal trajectories of the geodesics

$$\kappa_2 = (\kappa_g)_{u=\text{const}} = \frac{1}{\sqrt{G}} \frac{\partial}{\partial u} \sqrt{G} = \frac{G_u}{2G}, \qquad (3\text{–}5)$$

and for the Gaussian curvature of the surface

$$K = -\frac{1}{\sqrt{G}} \frac{\partial^2 \sqrt{G}}{\partial u^2} = \frac{(G_u)^2 - 2G G_{uu}}{4G^2}. \qquad (3\text{–}6)$$

This simple expression for K (rational in G and its derivatives) allows us to deduce some interesting interpretations of the Gaussian curvature. In the plane the ordinary system of polar coordinates

$$ds^2 = dr^2 + r^2\,d\varphi^2 = du^2 + u^2\,dv^2$$

is an example of geodesic polar coordinates. Here the conditions (3–4) are fulfilled. We now write, in accordance with this, for a system of geodesic polar coordinates on an arbitrary surface:

$$\sqrt{G(u,v)} = \sqrt{G(0,v)} + u\left[\frac{\partial}{\partial u}\sqrt{G(u,v)}\right]_{u=0} + \frac{u^2}{2}\left[\frac{\partial^2}{\partial u^2}\sqrt{G(u,v)}\right]_{u=0} + o(u^2)$$

or

$$\sqrt{G} = u + \alpha_1 u^2 + \alpha_2 u^3 + o(u^3),$$

where α_1 and α_2 are functions of v, which can be expressed in terms of the Gaussian curvature K_0 at $O(0,v)$ by means of Eq. (3–6) and its derivatives:

$$\frac{\partial^2\sqrt{G}}{\partial u^2} = -K_0\sqrt{G}; \qquad \frac{\partial^3\sqrt{G}}{\partial u^3} = -K_0\frac{\partial\sqrt{G}}{\partial u} - \sqrt{G}\left(\frac{\partial K}{\partial u}\right)_0.$$

Hence

$$2\alpha_1 = -K_0(u + \cdots)_0, \qquad 6\alpha_2 = -K_0 - \left(\frac{\partial K}{\partial u}\right)_0(u + \cdots)_0$$

or

$$\alpha_1 = 0, \qquad \alpha_2 = -\tfrac{1}{6}K_0,$$

and we obtain for $\sqrt{G}$ the formula

$$\sqrt{G} = u - \tfrac{1}{6}K_0 u^3 + R_1(u,v), \tag{3–7}$$

where $R_1(u,v)$ is of the order n, $n > 3$, in u. Hence

$$G = u^2 - \tfrac{1}{3}K_0 u^4 + R_2(u,v), \tag{3–8}$$

where $R_2(u,v)$ is of the order n, $n > 4$, in u.

Since the circumference C of a curve $u = u_0 = \text{constant}$ and the area A it encloses are given by

$$C = \int_0^{2\pi}\sqrt{G}\,dv, \qquad G = G(u_0,v),$$

$$A = \int_0^u\int_0^{2\pi}\sqrt{G}\,du\,dv, \qquad G = G(u,v),$$

we arrive at the following two expressions for K_0:

$$K_0 = \frac{3}{\pi} \lim_{u \to 0} \frac{2\pi u - C}{u^3},$$
$$K_0 = \frac{12}{\pi} \lim_{u \to 0} \frac{\pi u^2 - A}{u^4}.$$
(3–9)

The closed curve $u = u_0$ is the locus of all points at constant *geodesic distance* from a point (here O), that is, at constant distance measured along a geodesic. Such curves are called *geodesic circles*. The formulas (3–9) give a geometric illustration of the Gaussian curvature K as a bending invariant by comparing circumference and area of a small geodesic circle with the corresponding quantities for a plane circle of equal radius.

4–4 Geodesics as extremals of a variational problem. A popular (and the oldest) way to describe geodesics is to call them the curves of shortest distance between two points on a surface. This is certainly the case for the plane, where the geodesics are the straight lines, which are the shortest distances between any two of their points. But we must qualify this statement even for so simple a case as that of the sphere. For example, it is true that the shortest distance between two points P and Q on a sphere is along a geodesic, which on the sphere is a great circle. But there are two arcs of a great circle between two of their points, and only one of them is the curve of shortest distance, except when P and Q are at the end points of a diameter, when both arcs have the same length. This example of the sphere also shows that it is not always true that through two points only one geodesic passes: when P and Q are the end points of a diameter any great circle through P and Q is a geodesic and a solution of the problem of finding the shortest distance between two points.

Another simple example is offered on a cylinder, where two points on the same generator can be connected not only by means of the generator (the shortest distance), but also by an infinite number of helices of varying pitch, which wind around the cylinder and are all geodesics.

In order to find the shortest distance between P and Q, if such a one exists, we have to find the function $v = v(u)$ for which s obtains a minimum value. In order to find this function, let us suppose that the curve C solves the variational problem, so that the distance PQ, measured along C, is shorter than the distance PQ measured along any other curve C' between P and Q, when C' is taken close to C. Take C as the curve $v = 0$ of an orthogonal coordinate system

$$ds^2 = E\,du^2 + G\,dv^2,$$

and let the curve C' be determined by a small variation $\delta n = \sqrt{G}\,\delta v$ along the curves $u = $ constant, measured from C. If C' has the equation $v = v(u)$, then we take v so small that it can be identified with δv. When C' passes through $P(u = u_0)$ and $Q(u = u_1)$, then $v(u_0) = v(u_1) = 0$. The distances PQ along C and along C' are now

$$s = \int_{u_0}^{u_1} \sqrt{E}\,du, \qquad s + \delta s = \int_{u_0}^{u_1} \sqrt{E_1 + (v')^2 G}\,du,$$

respectively; here $E = E(u, 0)$, $E_1 = E(u, \delta v)$, $G = G(u, \delta v)$.

Since

$$E(u, \delta v) = E(u, 0) + \delta v E_v + \cdots, \qquad E_v = [\partial E(u, v)/\partial v]_{v=0},$$

we can write for the first variation δs of the arc length, neglecting all terms of order higher than δv (including $(v')^2$):

$$\begin{aligned}
\delta s &= \int_{u_0}^{u_1} \left[\sqrt{E}\left(1 + \delta v\,\frac{E_v}{E}\right)^{\frac{1}{2}} - \sqrt{E} \right] du \\
&= \int_{u_0}^{u_1} \left(\sqrt{E} + \frac{1}{2}\,\delta v\,\frac{E_v}{\sqrt{E}} - \sqrt{E} \right) du = \int_{u_0}^{u_1} \frac{1}{2}\,\delta v\,\frac{E_v}{\sqrt{E}}\,du \\
&= \int_{u_0}^{u_1} \frac{E_v}{2E\sqrt{G}}\,\delta n\,ds. \qquad (ds = \sqrt{E}\,du \text{ measured along } C)
\end{aligned}$$

Hence, according to Eq. (1–10)

$$\delta s = -\int_{u_0}^{u_1} \kappa_g\,\delta n\,ds, \tag{4–1}$$

where κ_g is the geodesic curvature along $C(v = 0)$.

A necessary condition that s be a minimum is that $\delta s = 0$ for all δn. This condition is fulfilled if the geodesic curvature κ_g vanishes along the curve C between P and Q. A necessary condition for the existence of a shortest distance between P and Q is that it be measured along a geodesic joining P and Q. In other words:

If there exists a curve of shortest distance between two points on a surface, then it is a geodesic.

We can also find this result by means of the general rules of the calculus of variations. They state that the curves which solve the variational problem

$$\delta \int f(x, y, y')\,dx = 0, \qquad y' = dy/dx,$$

are solutions of the *Euler-Lagrange equations,*

$$\frac{\partial f}{\partial y} - \frac{d}{dx}\frac{\partial f}{\partial y'} = 0.^*$$

Such curves are called the *extremals* of the variational problem. In the case of the geodesics, where

$$f(u, v, v') = \sqrt{E + 2Fv' + Gv'^2}, \qquad v' = dv/du,$$

the Euler-Lagrange equations are

$$\frac{E_v + 2F_v v' + G_v (v')^2}{\sqrt{E + 2Fv' + Gv'^2}} - \frac{d}{du}\frac{2(F + Gv')}{\sqrt{E + 2Fv' + G(v')^2}} = 0.$$

The left side of this equation can also be obtained if we substitute into Bonnet's formula for the geodesic curvature (Exercise 14, Section 4–2) the expressions

$$\varphi(u, v) = v - v(u), \qquad \varphi_u = -v', \qquad \varphi_v = 1, \qquad \frac{d}{du} = \frac{\partial}{\partial u} + v'\frac{\partial}{\partial v}.$$

This shows that the Euler-Lagrange equations of our problem can be written in the form $\kappa_g = 0$, so that the geodesics are the extremals of the variational problem $\delta s = 0$.

This result does not inform us when a given geodesic actually is a curve of shortest distance. A full discussion of this problem would lead us too far. A partial answer can be obtained if we return to our curve $C(v = 0)$ of Fig. 4–6, now a geodesic, and postulate that this curve is a part of a *field of geodesics,* at any rate in a region R close to C and including P and Q. Such a field is a one-parameter family (or *congruence*) of geodesics such that

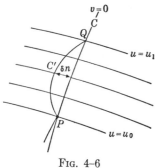

Fig. 4–6

there passes through every point of R one and only one geodesic of this family. We now take these geodesics as the curves $v = $ constant of our region and their orthogonal trajectories as the curves $u = $ constant. Then, varying again the curve $v = 0$ from C into C_1 between $P(u_0)$ and $Q(u_1)$, we have now $E = E_1 = 1$ and

$$\delta s = \int_{u_0}^{u_1}\sqrt{1 + Gv'^2}\,du - \int_{u_0}^{u_1}du > 0,$$

since $G = EG - F^2 > 0$. The arc length of C between P and Q is therefore smaller than any other curve C' between the same points, and we have found the theorem:

---

* See e.g. F. S. Woods, *Advanced calculus,* Ginn & Co., N. Y., 1934, Chap. 14.

If an arc of a geodesic can be im-bedded in a field of geodesics, then it offers the shortest distance between two of its points as compared to all other curves in the region for which the field is defined.

In other words, if there is but one geodesic arc between two points in a certain region, then that arc gives the shortest path in that region between these points.

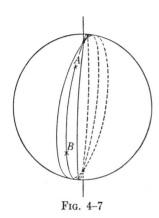

FIG. 4–7

This proposition can be illustrated on the sphere (Fig. 4–7). If we take on a great circle two points A and B which are not diametrically opposite, then we can imbed in a field of great circles that part of the great circle between A and B which is less than half the circumference, but the other half cannot be imbedded in this way. Indeed, any geodesic close to this other half passes through two dia-metrically opposite points of the sphere, which cannot happen in a field.

If the point Q moves on the geo-desic C away from P, then it may happen that it will reach a certain position where PQ will no longer be the shortest distance between P and Q. Suppose that this happens at point R. Then, if the geodesics through P near C have an envelope E, it can be shown that R is the point where C is tangent to E for the first time. This point R is called *conjugate* to P on the geodesic C. An example of such geodesics is offered by the ellipsoid (see Fig. 4–8, where we see the projection on the plane of symmetry).

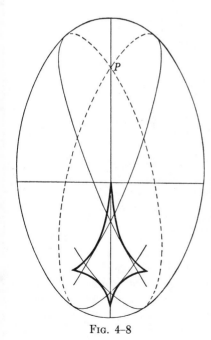

FIG. 4–8

Further discussion of this case is difficult without full use of the calculus of variations. However, we can show by simple reasoning that *conjugate points do not exist on regions of a surface where K is negative*. For this purpose let us consider a geodesic C, and now consider it as the curve $u = 0$ of a geodesic coordinate system

$$ds^2 = du^2 + G\,dv^2.$$

The distance between two points $P(v = v_1)$, $Q(v = v_2)$ on C, measured along C, and measured along another curve $u = u(v)$ passing through P and Q is

$$s = \int_{v_0}^{v_1} \sqrt{G}\,dv, \qquad s + \delta s = \int_{v_0}^{v_1} \sqrt{u'^2 + G}\,dv, \qquad u' = du/dv,$$

respectively; we take u small and u' finite. By a reasoning similar to the one used in the case of geodesic polar coordinates we find for $\sqrt{G}$ a series expansion of the form

$$\sqrt{G} = 1 - \tfrac{1}{2}Ku^2 + o(u^2), \qquad K = K(0, v),$$

so that

$$s + \delta s = \int_{v_0}^{v_1} \sqrt{u'^2 + (1 - Ku^2 + o(u^2))}\,dv = \int_0^{v_1} [1 + \tfrac{1}{2}(u'^2 - Ku^2) + \epsilon]\,dv,$$

where ϵ is small compared to u^2. Hence

$$\delta s = \tfrac{1}{2}\int_{v_0}^{v_1} (u'^2 - Ku^2 + \epsilon)\,dv.$$

For $K < 0$ we can always keep $\delta s > 0$ provided we stay in a sufficiently small neighborhood of C.

The question whether a geodesic is the shortest distance between two points was first raised by C. G. J. Jacobi in his *Vorlesungen über Dynamik* (1842/43), see his *Ges. Werke*). The theory is discussed in G. Darboux's *Leçons III*, Ch. V, and in books on the calculus of variations. Fig. 4–8 is taken from A. von Braunmühl, *Mathem. Annalen* **14** (1879), p. 557.

4–5 Surfaces of constant curvature. We have seen that surfaces of zero Gaussian curvature are identical with developable surfaces (Section 2–8). The simplest example of a surface of constant non-zero Gaussian curvature (or *surface of constant curvature*) is a sphere, where $K = a^{-2}$ (a the radius) and is therefore positive. It can easily be seen that the sphere is not the only surface with this property. We have only to take a cap of a sphere (as usual we discuss here properties holding for certain segments of a surface, not necessarily for surfaces as a whole) and think of it as consisting of some inelastic material, perhaps thin brass. Then it is possible to give this piece of brass all kinds of shapes, and all surfaces thus formed have the same constant curvature as the spherical cap.

This by itself does not prove that all surfaces of the same constant curvature (or at any rate certain regions on it) can be obtained from each other by bending. But we can prove here a theorem giving at least an approach to this subject. For this purpose *we define an isometric correspondence between two surfaces as a one-to-one point correspondence such that at corresponding points the ds^2 are equal.* We say that the surfaces are *isometrically mapped* upon each other. Bending is a way of obtaining isometric mapping, but an isometric correspondence need not be the result of bending. We can now prove *Minding's theorem:*

All surfaces of the same constant curvature are isometric.

We distinguish among the cases $K > 0$, $K = 0$, $K < 0$.

I. Let $K = 0$. Take a system of geodesic polar coordinates

$$ds^2 = du^2 + G\,dv^2. \tag{5-1}$$

Since K satisfies Eq. (3–6), the differential relation

$$K = -\frac{1}{\sqrt{G}}\frac{\partial^2 \sqrt{G}}{\partial u^2}, \tag{5-2}$$

we find that $\sqrt{G}$ is of the form

$$\sqrt{G} = u c_1(v) + c_2(v). \tag{5-3}$$

We can impose on G the conditions (compare with Section 4–3)

$$(\sqrt{G})_{u=0} = 0, \qquad (\partial\sqrt{G}/\partial u)_{u=0} = 1. \tag{5-4}$$

Hence $\sqrt{G} = u$ and

$$ds^2 = du^2 + u^2\,dv^2.$$

This expression for ds^2 can be obtained for all surfaces with $K = 0$ by taking on it a geodesic polar coordinate system. This can be done in ∞^3 ways (∞^2 possible points for origin, ∞^1 choices of $v = 0$). All surfaces of zero curvature therefore are isometric. Taking $x = u \cos v$, $y = u \sin v$, we obtain ds^2 in the form

$$ds^2 = dx^2 + dy^2, \tag{5-5}$$

which shows that all developable surfaces can be isometrically mapped on a plane, the curvilinear coordinates corresponding to the rectangular cartesian coordinates in the plane.

This isometrical mapping can actually be accomplished by bending. To show this, let us take a tangential developable with the equation (Section 2–3)

$$\mathbf{y} = \mathbf{x}(s) + v\mathbf{t}(s), \qquad \mathbf{y}_s = \mathbf{t} + \kappa v\mathbf{n}, \qquad \mathbf{y}_v = \mathbf{t}$$

and

$$ds^2 = (1 + \kappa^2 v^2)\, ds^2 + 2\, dv\, ds + dv^2. \tag{5–6}$$

The shape of the curve $\mathbf{x} = \mathbf{x}(s)$ is fully determined by its $\kappa(s)$ and $\tau(s)$. Let us now consider a family of curves given by $\kappa = \kappa(s)$ and $\tau_1 = \mu\tau(s)$, where μ takes all values from $\mu = 1$ to $\mu = 0$. For $\mu = 0$ we obtain a plane curve. When μ passes to all values from $\mu = 1$ to $\mu = 0$, the corresponding tangential developable maintains its ds^2 in the form (5–6), since $\kappa(s)$ does not change. Thus the surface passes through a set of continuous isometric transformations from (5–6) to the case $\tau = 0$, which is a plane. Then the tangents to the space curve have become the tangents to the plane curve with the same (first) curvature. In other words (see the remark made in Section 2–4 (Fig. 2–12)), *when a piece of paper with a curve C on it is bent, and C becomes the edge of regression C_1 of a developable surface, then C_1 passes into a space curve with the same (first) curvature as C.*

II. Let $K > 0$. Again take a system (5–1) of geodesic polar coordinates. Now $\sqrt{G}$ is a solution of

$$\partial^2\sqrt{G}/\partial u^2 + K\sqrt{G} = 0. \tag{5–7}$$

This is a linear differential equation with constant coefficients; hence

$$\sqrt{G} = c_1 \sin \sqrt{K}\,(u + c_2), \tag{5–8}$$

where c_1 and c_2 are functions of v. If we impose upon $\sqrt{G}$ the conditions (5–4) we obtain

$$\sqrt{G} = \frac{1}{\sqrt{K}} \sin u\sqrt{K} \tag{5–9}$$

and

$$ds^2 = du^2 + \frac{1}{K}\,(\sin^2 \sqrt{K}\,u)\, dv^2. \tag{5–10}$$

Reasoning as in the case $K = 0$, we see that we can find, on every surface with the same K, and in ∞^3 ways, a coordinate system in which ds^2 takes the form (5–10), and the isometric mapping is accomplished. In other words:

Every surface of constant positive curvature K can be isometrically mapped on a sphere of radius $K^{-\frac{1}{2}}$. This theorem only holds, of course, for such regions of the surface where we can introduce a system (5–1); those regions can then be isometrically mapped on corresponding regions of a sphere.

III. Let $K < 0$. Reasoning in the same way as before, we obtain the equation

$$\partial^2 \sqrt{G}/\partial u^2 - L\sqrt{G} = 0, \qquad L = -K, \tag{5-11}$$

which, under the conditions (5–4), has the solution

$$\sqrt{G} = \frac{1}{\sqrt{L}} \sinh u\sqrt{L},$$

so that

$$ds^2 = du^2 + \frac{1}{L} (\sinh u\sqrt{L})^2 \, dv^2, \tag{5-12}$$

which shows that here again all surfaces of this type can be isometrically mapped on each other, and in ∞^3 ways.

This theory is due to Minding, *Journal für Mathem.* **19**, 1839, and was further elaborated by many other mathematicians, notably by E. Beltrami. Excellent expositions of the theory of surfaces of constant curvature can be found in Darboux's *Leçons* or Bianchi's *Lezioni*. Hilbert, loc. cit., Section 3–5, also proved that there does not exist a surface of constant negative curvature which is without singularities and everywhere analytical.

4-6 Rotation surfaces of constant curvature. There is a great variety of surfaces of constant curvature, which can be gathered from the fact that all the many varieties of developable surfaces are surfaces of zero curvature. The best known types of surfaces of constant curvature $K \neq 0$ are the rotation surfaces of constant curvature. We shall here give an outline of their theory.

Such surfaces can be given by the parametric representation (Section 2–2),

$$x = r \cos \varphi, \qquad y = r \sin \varphi, \qquad z = f(r), \tag{6-1}$$

with corresponding first fundamental form

$$ds^2 = (1 + f'^2) \, dr^2 + r^2 \, d\varphi^2. \tag{6-2}$$

By means of the transformation

$$du = \sqrt{1 + f'^2} \, dr, \tag{6-3}$$

Eq. (6–2) can be brought into the geodesic form

$$ds^2 = du^2 + G \, d\varphi^2,$$
$$G = G(u) = r^2. \tag{6-4}$$

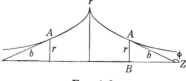

Fig. 4–9

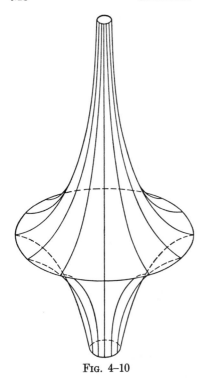

FIG. 4–10

The function G has the special property that it depends only on u, and it can be readily shown (see Eq. (6–7)) that *this property is characteristic of all surfaces isometric with rotation surfaces*. In this case the Eq. (5–7) takes the form

$$d^2\sqrt{G}/du^2 + K\sqrt{G} = 0$$
$$\text{(with straight } d\text{).}$$

The solution for $K = a^{-2} > 0$ is given by

$$\sqrt{G} = c_1 \cos u/a + c_2 \sin u/a,$$
$$c_1, c_2 \text{ constants,} \quad (6\text{–}5)$$

and for $K < 0$, $K = -b^{-2}$ it is given by

$$\sqrt{G} = c_1 e^{u/b} + c_2 e^{-u/b},$$
$$c_1, c_2 \text{ constants.} \quad (6\text{–}6)$$

The c_1 and c_2 can always be selected in accordance with conditions (5–4), but sometimes it is more convenient to select them otherwise. For the moment we leave them undetermined. Since according to (6–2), (6–3), and (6–4):

$$r = \sqrt{G}, \qquad z = f(r) = \int \sqrt{1 - \left(\frac{dr}{du}\right)^2}\, du, \qquad (6\text{–}7)$$

we find for the profile of the rotation surfaces of constant curvature: (a) for $K = a^{-2} > 0$: $r = c_1 \cos u/a + c_2 \sin u/a$,

$$z = \int \sqrt{1 - a^{-2}(-c_1 \sin u/a + c_2 \cos u/a)^2}\, du,$$

(b) for $K = -b^{-2} < 0$: $r = c_1 \exp. (u/b) + c_2 \exp. (-u/b)$,

$$z = \int \sqrt{1 - b^{-2}[c_1 \exp. (u/b) - c_2 \exp. (-u/b)]^2}\, du,$$

where the choice of c_1 and c_2 may result in different types of surfaces.

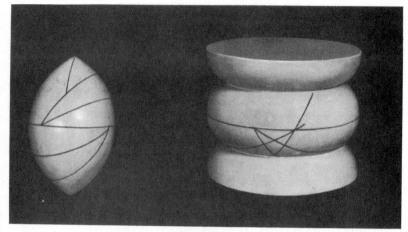

Fig. 4–11

These equations show that, in general, *the determination of rotation surfaces of constant curvature leads to elliptic integrals*. However, there are special cases in which these integrals may be evaluated in terms of elementary functions. One case is that in which $K = a^{-2}$, $c_1 = a$, $c_2 = 0$. Then $ds^2 = du^2 + a^2(\cos^2 u/a) \, dv^2$,

$$r = a \cos u/a, \qquad z = a \sin u/a + \text{constant},$$

and we obtain the sphere (also obtained for $c_2 = a$, $c_1 = 0$). Another case is that in which $K = -b^{-2}$, $c_1 = b$, $c_2 = 0$. Then $ds^2 = du^2 + b^2 e^{2u/b} \, dv^2$, and

$$r = b \exp. (u/b), \qquad z = \int \sqrt{1 - \exp. (2u/b)} \, du. \qquad (6\text{–}8)$$

Instead of evaluating the integral, we observe that here

$$\frac{dr}{dz} = \frac{r}{\sqrt{b^2 - r^2}}, \qquad (6\text{–}9)$$

which shows that the curve $r = f(z)$ has the property that the segment of the tangent between the point of tangency A and the point of intersection with the Z-axis is constant and $= b$ (both $\pm \sqrt{b^2 - r^2}$ give the same curve, Fig. 4–9). This curve is the *tractrix*, and the corresponding rotation surface is called the *pseudosphere*, with *pseudoradius* b. We have thus found the following result:

Every surface of constant negative curvature $K = -b^2$ can be isometrically mapped on a pseudosphere of pseudoradius b.

Fig. 4–12

The pseudosphere has a singular central circle and thins out asymptotically on both ends (Fig. 4–10).

We show in Fig. 4–11 some rotation surfaces of constant positive curvature, and in Fig. 4–12 some rotation surfaces of constant negative curvature. Their profiles can be expressed with the aid of elliptic functions.

4–7 Non-Euclidean geometry. The geometry of surfaces of constant curvature receives additional interest through its relation with the so-called non-Euclidean geometry. This geometry was the outcome of the century-long quest for a proof of the statement, introduced in Euclid's *Elements* (c. 300 B.C.) as the fifth postulate:

"If a straight line falling on two straight lines in the plane makes the interior angles on the same side less than two right angles, then the two straight lines, if produced indefinitely, will meet on that side on which are the angles less than two right angles."

This postulate is equivalent to another one, known as the parallel axiom.* This asserts that through a point outside of a line one and only

---

* The precise difference between "postulates" and "axioms" in Euclid is not clear.

one parallel to that line can be drawn. From it we can deduce that the sum of the angles of a triangle is equal to two right angles. Both fifth postulate and parallel axiom are far from self-evident, so that already in Antiquity attempts were made to prove them, that is, to obtain them by logical deduction from other postulates or axioms considered more immediately evident (such as the axiom that two points determine a line). Though in Antiquity and all through modern times until c. 1830 several results were obtained which enriched our understanding of the axiomatic structure of geometry, a proof of the parallel axiom was never found. Many so-called demonstrations smuggled in some unavowed assumptions. Some mathematicians — the more penetrating thinkers — derived from the denial of the parallel axioms some startling conclusions which they considered absurd, but what seems "absurd" to one generation may still be logical and acceptable to another. Per-

haps the most valuable contribution to this type of literature was made by the Italian Jesuit G. Saccheri (*"Euclid freed from every birthmark,"* 1733, Latin). Saccheri considered a quadrangle of which two opposite sides *AB, CD* are equal and perpendicular to the third side *BC* (Fig. 4–13). Then he considered three possibilities:

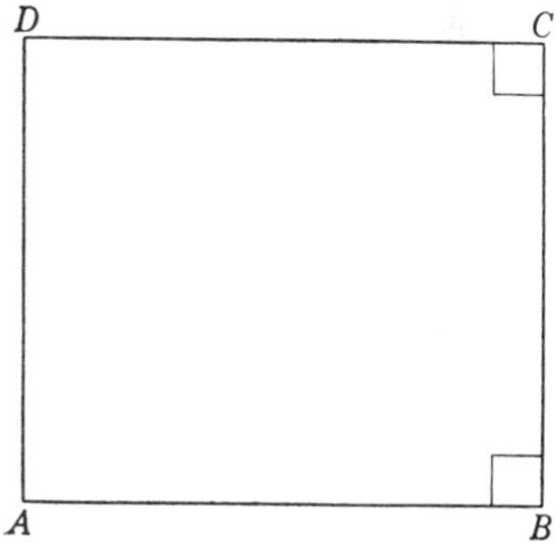

FIG. 4–13

(a) The angles at *A* and *D* are obtuse ("obtuse angle hypothesis").

(b) These angles are right angles ("right angle hypothesis").

(c) These angles are acute ("acute angle hypothesis").

He then proved that when one hypothesis is accepted for one quadrangle, it must be accepted for all quadrangles. From this follows again that the sum of the angles of a triangle is greater than, equal to, or smaller than two right angles provided hypothesis (a), (b), or (c), respectively, holds. Saccheri's work remained little known, but several other mathematicians came to related results, notably Legendre. Saccheri and Legendre rejected the hypothesis of the obtuse angle because it leads to an absurdity if we assume that a line has infinite length; they also had objections to the acute angle hypothesis.

Gauss seems to have been the first to believe in the independence of the parallel axiom, so that he accepted the logical possibility of a geometry in which the parallel axiom is replaced by another one, such as the hypothesis of the acute angle. Gauss believed that the geometry of space can be found by means of experiments, hereby opposing a favorite tenet of the

Kantian philosophy prevalent in his days. The first mathematician to state publicly (1826) that a geometry can be construed independently of the parallel axiom was N. I. Lobačevskiĭ, professor at Kazan, whose first work on this subject was published in 1829. He was followed independently by the young Hungarian officer John Bolyai (Bolyai Janos), whose essay was published as an Appendix to his father's big tome on geometry (1832). Gauss commended this paper, writing that he could not praise it, as this would mean praising himself, since he had come to the same conclusions — a kind of praise eminently fitted to discourage the ambitious young officer.

Both Lobačevskiĭ's and Bolyai's geometry are based on the hypothesis of the acute angle. In this case the lines through a point P in a plane can be divided into two classes, those that intersect a given line l not passing through P, and those that do not intersect it. These classes are separated by two lines which can be called parallel to l through P. In this geometry as well as in that of Euclid a straight line is an open line of infinite length. If we postulate that straight lines are closed and have a finite length, we can construct a geometry based on the hypothesis of the obtuse angle; this geometry was sketched by B. Riemann (1854) inside the framework of the more general so-called Riemannian geometries. Under the obtuse angle hypothesis all lines passing through a point P outside a line l in a plane intersect l. Thus all three assumptions of Saccheri could now be introduced as the bases of possible geometries, which became known as Euclidean and non-Euclidean geometries.

However, most mathematicians continued to believe that although there were as yet no logical inconsistencies discovered in the non-Euclidean geometries, it might very well be possible to find such inconsistencies after a more penetrating study of the properties of these geometries. Consequently, little attention was paid to them despite the authority of Riemann. This situation was radically changed through a paper by the Italian mathematician E. Beltrami, *An attempt to interpret the non-Euclidean geometry* (1868). Beltrami proved that when we take the geodesics of a surface of constant negative curvature and interpret them as "lines," interpreting angles and lengths according to the ordinary methods of differential geometry, then we obtain a geometry in which the hypothesis of the acute angle holds. The whole geometry of Lobačevskiĭ-Bolyai could thus be interpreted on a surface of constant negative curvature, parallel lines becoming asymptotic geodesics. Thus Beltrami proved that the consistency of Euclidean geometry implied the consistency of Lobačevskiĭ-Bolyai geometry, since an inconsistency in the latter could be interpreted as an inconsistency in the theory of surfaces of constant negative curvature, which itself is based on Euclidean postulates. Every inconsistency in one geometry implied an inconsistency in the other. Beltrami's mapping of the one geometry on the other was followed by several other such representations, not only for the case of the acute angle hypothesis, but also for the hypothesis of the obtuse angle. We now know that Riemann's geometry can be interpreted on a sphere by taking the great circles as "straight lines," provided we identify two diametrically opposite points (otherwise two points would not always determine one and only one line). The logical equivalence of the three geometries has thus been fully established.

More details can be found in books on non-Euclidean geometry, notably in J. L. Coolidge, *The elements of non-Euclidean geometry*, Oxford, 1909, or R. Bonola, *Non-Euclidean geometry*, Chicago, 2d ed., 1938. A neat summary is found in J. L. Coolidge, *A history of geometrical methods*, Oxford, 1940, pp. 68–97. English translations of Lobačevskiĭ and Bolyai's papers were published by G. B. Halsted: *Geometrical researches on the theory of parallels*, Chicago, London, 1914, 50 pp.; *New principles of geometry*, Austin, Tex., 1897, 27 pp. (both by Lobačevskiĭ); *The science absolute of space*, Chicago, London, 1914, xxx, + 71 pp. (by Bolyai). Beltrami's *Saggio di interpretazione della geometria non-euclidea* is found in Opere I, pp. 374–405. There exists a French translation: *Annales Ecole Normale* 6, 1869, pp. 251–288. See also H. S. M. Coxeter, *Non-Euclidean geometry*, Toronto, 2d ed., 1947.

It should be stressed that Beltrami's method only allows us to map a part of the Lobačevskiĭ plane on a part of a surface of negative curvature. Hilbert has shown (see Section 3–5) that it is impossible to continue an analytical surface of constant negative curvature indefinitely without meeting singular lines (at any rate when this surface lies in ordinary Euclidean space, see also Section 4–5).

4–8 The Gauss-Bonnet theorem. There exists a classical theorem, first published by Bonnet in 1848, but which probably was already known to Gauss, which we shall prove as an example of the differential geometry of surfaces in the large. It is an application of Green's theorem, known from the theory of line integrals and surface integrals in the plane, to the integral of the geodesic curvature.

Green's theorem asserts that if P and Q are any two functions of x and y in the plane for which the partial derivatives $\partial P/\partial y$ and $\partial Q/\partial x$ are continuous throughout a certain region R with area A bounded by a closed piecewise smooth curve C, then (Fig. 4–14), the sense of C being counterclockwise,

$$\int_C P\,dx + Q\,dy = \iint_A \left(\frac{\partial Q}{\partial x} - \frac{\partial P}{\partial y}\right) dx\,dy.$$

The region R may be the sum of a finite number of areas, each of which is such that any straight line through any interior point of the region cuts the boundary in exactly two points. The theorem holds equally on a surface if x and y are replaced by the curvilinear coordinates u and v, the parametric lines covering R in

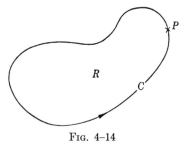

Fig. 4–14

such a way that through every point inside R passes exactly one curve $u = $ constant and one curve $v = $ constant.

If $P(u, v)$ and $Q(u, v)$ are two functions of u and v on a surface, then, according to Green's theorem and the expression in Chapter 2, Eq. (3–4) for the element of area:

$$\int_C P\,du + Q\,dv = \iint_A \left(\frac{\partial Q}{\partial u} - \frac{\partial P}{\partial v}\right) \frac{1}{\sqrt{EG - F^2}}\,dA,$$

where dA is the element of area of the region R enclosed by the curve C. With the aid of this theorem we shall evaluate

$$\int_C \kappa_g\,ds,$$

where κ_g is the geodesic curvature of the curve C. If C at a point P makes the angle θ with the coordinate curve $v = $ constant and if the coordinate curves are orthogonal, then, according to Liouville's formula (1–13):

$$\kappa_g\,ds = d\theta + \kappa_1(\cos\theta)\,ds + \kappa_2(\sin\theta)\,ds.$$

Here κ_1 and κ_2 are the geodesic curvatures of the curves $v = $ constant and $u = $ constant respectively. Since

$$\cos\theta\,ds = \sqrt{E}\,du. \qquad \sin\theta\,ds = \sqrt{G}\,dv,$$

we find by application of Green's theorem:

$$\int_C \kappa_g\,ds = \int_C d\theta + \iint_A \left(\frac{\partial}{\partial u}\left(\kappa_2\sqrt{G}\right) - \frac{\partial}{\partial v}\left(\kappa_1\sqrt{E}\right)\right) du\,dv.$$

The Gaussian curvature can be written, according to Chapter 3, Eq. (3–7),

$$K = -\frac{1}{2\sqrt{EG}}\left[\frac{\partial}{\partial u}\frac{G_u}{\sqrt{EG}} + \frac{\partial}{\partial v}\frac{E_v}{\sqrt{EG}}\right]$$

$$= \frac{1}{\sqrt{EG}}\left[-\frac{\partial}{\partial u}\left(\kappa_2\sqrt{G}\right) + \frac{\partial}{\partial v}\left(\kappa_1\sqrt{E}\right)\right],$$

so that we obtain as a result the formula

$$\int_C \kappa_g\, ds = \int_C d\theta - \iint_A K\, dA. \tag{8–1}$$

The integral $\iint_A K\, dA$ is known as the *total* or *integral curvature*, or *curvature integra*, of the region R, the name by which Gauss introduced it.

Let us first take a smooth curve C, as the boundary of region R on which, therefore, there are no points where the slope has discontinuities. Then we can contract C continuously without changing $\int_C d\theta$, since this is an integral multiple of 2π. Let A be a simply connected region, that is, a region which by continuous contraction of C can be reduced to a point. When C is reduced to approach a point then $\int_C d\theta = 2\pi$, and we have found the theorem that

$$\int_C \kappa_g\, ds + \iint_A K\, dA = 2\pi. \tag{8–2}$$

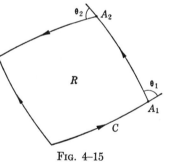

FIG. 4–15

When C consists of k arcs of smooth curves (Fig. 4–15) making exterior angles $\theta_1, \theta_2, \ldots, \theta_k$ at the vertices, $A_1, A_2, \ldots, A_k$, where the arcs meet, then we must keep in mind that $d\theta$ in Eq. (8–1) measures only the change of θ along the smooth arcs, where we measure $\int \kappa_g\, ds$, and not the jumps at the vertices. The total change in θ along C is still 2π, but only part of it is due to the change of θ along the arcs. The remainder is due to the angles $\theta_1, \theta_2, \ldots \theta_k$. For the sum of the line and area integral we therefore get $2\pi - \theta_1 - \theta_2 - \cdots - \theta_k$. This result is expressed in the *Gauss-Bonnet theorem*:

If the Gaussian curvature K of a surface is continuous in a simply connected region R bounded by a closed curve C composed of k smooth arcs making *at the vertices exterior angles* $\theta_1, \theta_2, \ldots \theta_k$, *then:*

$$\boxed{\int_C \kappa_g \, ds + \iint_A K \, dA = 2\pi - \sum_i \theta_i,} \quad i = 1, 2, \ldots k$$

where κ_g *represents the geodesic curvature of the arcs.*

This theorem was first published by O. Bonnet in the previously mentioned paper in the *Journ. Ecole Polytechnique* **19** (1848), pp. 1–146, as a generalization of Gauss' theorem on a geodesic triangle, see application II of this section. Bonnet proved it first for geodesic triangles, using the method of our Exercise 6, this section. The relation of this theorem to Green's theorem was clarified by G. Darboux, *Leçons III*, p. 122, etc.

The integral curvature $\iint K \, dA$ has a simple geometrical interpretation, by means of which Gauss originally introduced it. When we draw through a point O lines parallel to the surface normals at the points of a region R on a surface S bounded by an arc C, then they intersect the unit sphere in a region R_1 bounded by an arc C_1 which is the *spherical image* of C (see Exercise 6, Section 2–11). We can introduce coordinates (u, v) on the unit sphere such that a point and its spherical image have the same u and v. The equation of C_1 is $\mathbf{N} = \mathbf{N}(s)$, where s is the arc length of C. The element of arc $d\sigma$ of C_1 is given by $d\sigma^2 = d\mathbf{N} \cdot d\mathbf{N}$, and the element of area dA_s of R_1, according to Chapter 2, Eq. (3–4), by

$$dA_s = \sqrt{(\mathbf{N}_u \cdot \mathbf{N}_u)(\mathbf{N}_v \cdot \mathbf{N}_v) - (\mathbf{N}_u \cdot \mathbf{N}_v)^2} \, du \, dv$$
$$= \sqrt{(\mathbf{N}_u \times \mathbf{N}_v) \cdot (\mathbf{N}_u \times \mathbf{N}_v)} \, du \, dv.$$

According to the Weingarten formulas, Chapter 3, Eq. (2–9):

$$\mathbf{N}_u \times \mathbf{N}_v = \frac{(Ff - Ge)(Ff - Eg) - (Fe - Ef)(Fg - Gf)}{(EG - F^2)^2} \mathbf{x}_u \times \mathbf{x}_v$$
$$= \frac{eg - f^2}{EG - F^2} \mathbf{x}_u \times \mathbf{x}_v = K\mathbf{x}_u \times \mathbf{x}_v.$$

Hence

$$dA_s = |K| \sqrt{(\mathbf{x}_u \times \mathbf{x}_v) \cdot (\mathbf{x}_u \times \mathbf{x}_v)} \, du \, dv = |K| \sqrt{EG - F^2} \, du \, dv = |K| \, dA,$$

so that *the integral (absolute) curvature of a region on the surface is equal to the area of its spherical image.* This was the property stressed by Gauss.

This property of the integral curvature was already known to the French school of Monge before Gauss pointed out its significance for the intrinsic geometry of a surface, see O. Rodrigues, loc. cit. Section 2–9.

When the area becomes smaller and smaller, we obtain the property that, with the appropriate sign,

$$K = \lim_{\Delta A \to 0} \frac{\Delta A_s}{\Delta A}.$$

The Gaussian curvature is the limit of the ratio of the small areas on the spherical image and on the surface which correspond to each other, when the areas become smaller and smaller.

Here we must take into consideration that for a hyperbolic point the contour of ΔA_s is traversed in a direction opposite to that of ΔA.

This definition of curvature corresponds to that given by $\kappa = d\varphi/ds$ for plane curves.

We shall finally give some applications of the theorem of Gauss-Bonnet.

I. When we apply this theorem to a triangle formed by an arc of a curve between two neighboring points P and Q and the geodesics tangent to C at P and Q (Fig. 4–16) we find that $\iint K \, dA$ is of order $(\Delta\varphi)^3$, so that but for quantities of order $(\Delta\varphi)^2$ and higher (where $\Delta\varphi$ is the angle between the tangent geodesics at P and Q):

$$\int_P^Q \kappa_g \, ds = 2\pi - (-\Delta\varphi) - \pi - \pi = \Delta\varphi,$$

where $\Delta\varphi$ is the angle between the geodesics. When Q approaches P we find

$$\boxed{\kappa_g = d\varphi/ds,} \tag{8–3}$$

which formula generalizes for surfaces the well-known formula for the ordinary curvature of a plane curve and gives us at last a simple definition of the geodesic curvature in terms of quantities *on* the surface.

II. When the boundary C consists of k geodesic lines, intersecting at interior angles $\alpha_1, \alpha_2, \ldots, \alpha_k, \alpha_i = \pi - \theta_i, i = 1, \ldots k$, we have a *geodesic polygon*. The integrals of the geodesic curvature vanish and we obtain for the integral curvature:

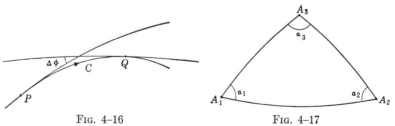

Fig. 4–16   Fig. 4–17

$$\iint_A K \, dA = 2\pi - k\pi + \sum \alpha_i = \sum \alpha_i - (k-2)\pi.$$

When $k = 3$ we find the theorem (Fig. 4–17):

The integral curvature of a geodesic triangle is equal to the excess of the sum of its angles over π radians. Gauss wrote: "this theorem, if we mistake not, ought to be counted among the most elegant in the theory of curved surfaces."

This theorem is of particular interest for a surface of constant curvature. In this case the integral curvature is equal to KA, where A is the area of the geodesic polygon. We express this result in the theorem:

The area of a geodesic triangle on a surface of constant curvature is equal to the quotient of $(\alpha_1 + \alpha_2 + \alpha_3 - \pi)$ and K, where the α are the angles of the triangle.

This result generalizes the well-known *theorem of Legendre* for a sphere, where $\alpha_1 + \alpha_2 + \alpha_3 - \pi$ is the *spherical excess:*

The area of a spherical triangle is equal to the product of its spherical excess into the square of the radius.

It also shows that the sum of the angles of a geodesic triangle is greater than π for a surface of constant positive curvature and less than π for a surface of constant negative curvature. For developable surfaces the sum is π.

III. A surface which can be brought into one-to-one continuous correspondence with the sphere is called *topologically equivalent* to the sphere. Such surfaces can be obtained by continuous deformation of a sphere without breaking it and without bringing separated parts of its surface together. Such a surface can be separated into two simply connected regions I and II by a closed smooth curve C without double points (Fig. 4–18). If we apply the Gauss-Bonnet theorem to such a curve, first as boundary of region I, and then of region II, we obtain, if we traverse C both times in such direction that the enclosed region stays on the left side:

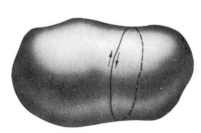

$$\int_C \kappa_g \, ds + \iint_I K \, dA = 2\pi,$$

$$\int_C \kappa_g \, ds + \iint_{II} K \, dA = 2\pi.$$

Adding these two equations, we obtain the result

$$\iint K \, dA = 4\pi.$$

Fig. 4–18

FIG. 4–19

In words:

The integral curvature of a surface which is topologically equivalent to the sphere is 4π.

This theorem characterizes the integral curvature as a *topological invariant.* If we apply a similar reasoning to a surface which is topologically equivalent to a torus, then we see that it can be changed into a simply connected region R by two cuts (Fig. 4–19). These cuts can be so applied that this region R is bounded by four smooth curves intersecting at $\pi/2$ radians. When we apply the Gauss-Bonnet theorem to this boundary, we obtain

$$\int_C \kappa_g \, ds + \iint_A K \, dA = 2\pi - 4 \cdot \frac{\pi}{2} = 0.$$

The integral of the geodesic curvature vanishes, and

$$\iint K \, dA = 0.$$

The integral curvature of a surface which is topologically equivalent to a torus is zero.

EXERCISES

1. Integrate the differential equation (6–9) of the tractrix. Write $r = b \sin \varphi$.

2. The geodesic curvature of a curve C on a surface S is equal to the ordinary curvature of the plane curve into which C is deformed when the developable surface enveloped by the tangent planes to S along C is rolled out on a plane.

3. A geodesic C on S is also a geodesic on the developable surface enveloped by the tangent planes to S along C.

4. When a geodesic makes the angle θ with the curves $v =$ constant of a geodesic coordinate system for which $ds^2 = du^2 + G \, dv^2$, then

$$d\theta + (\partial\sqrt{G}/\partial u) \, dv = 0.$$

5. If two families of geodesics intersect at constant angle, the surface is developable.

6. Prove Gauss' theorem on geodesic triangles by taking two sides as the coordinate curves $v = 0$ and $v = v_0$ of a geodesic coordinate system.   This was Gauss' proof of his theorem.   Hint: Use the result of Exercise 4.

7. Show that for a surface of negative curvature the hypothesis of the acute angle holds for quadrangles formed by geodesics (see Section 3–7).

8. The integral curvature of a closed surface of *genus p* (or *connectivity* $2p + 1$), that is, a surface which can be made into a simply connected region by $2p$ simple closed cuts (e.g., a sphere with p handles) is $4\pi(1 - p)$.

9. Prove Euler's theorem for convex polyhedrons $F - E + V = 2$ (F = number of faces, E = number of edges, V = number of vertices) by applying the Gauss-Bonnet theorem (W. Blaschke, *Differentialgeometrie I*, p. 166).

10. By using the Gauss-Bonnet theorem, show that two geodesics on a surface of negative curvature cannot meet in two points and enclose a simply connected area.   (Compare the theorem on p. 144.)

11. *Differential parameters.*   E. Beltrami, in 1864–1865, introduced the following expressions, in which u and v are curvilinear coordinates on a surface:

$$\nabla(\varphi, \psi) = \frac{E\varphi_v\psi_v - F(\varphi_u\psi_v + \varphi_v\psi_u) + G\varphi_u\psi_u}{EG - F^2},$$

$$\Delta\varphi = \frac{1}{\sqrt{EG - F^2}} \left\{ \frac{\partial}{\partial u} \frac{G\varphi_u - F\varphi_v}{\sqrt{EG - F^2}} + \frac{\partial}{\partial v} \frac{E\varphi_v - F\varphi_u}{\sqrt{EG - F^2}} \right\},$$

which he called the *first* and *second differential parameter* of the functions ψ and φ of u, v, respectively.   Show

(a) that the derivative $d\varphi/ds$ of a function $\varphi(u, v)$ on the surface in the direction ds is maximum when $(d\varphi/ds)^2 = \nabla(\varphi, \varphi)$,

(b) that for the plane and the notation of vector analysis

$$\nabla(\varphi, \psi) = (\text{grad } \varphi) \cdot (\text{grad } \psi), \qquad \Delta\varphi = \text{div grad } \varphi = \text{Lap } \varphi,$$

(c) that $\pm \kappa_g = \dfrac{\Delta\varphi}{\sqrt{\nabla(\varphi, \varphi)}} + \nabla\left(\varphi, \dfrac{1}{\sqrt{\nabla(\varphi, \varphi)}}\right).$

(Beltrami, *Opere mat. I*, pp. 107–198.   Lamé, in the *Leçons sur les coordonnées curvilignes*, 1859, already had used differential parameters for the plane and for space.)

12. Show that when two independent systems of geodesic parallels are selected as parametric curves, the element of arc can be written

$$ds^2 = \csc^2 \omega(du^2 + 2 \cos \omega \, du \, dv + dv^2),$$

where ω is the angle between the parametric curves.

13. *Geodesic ellipses and hyperbolas.*   The locus of a point for which $u + v =$ constant in the coordinate system of Exercise 12 is called a *geodesic ellipse*.   Simi-

larly, we call the locus for which $u - v =$ constant a *geodesic hyperbola*. Show that ds^2 can be written

$$ds^2 = (\csc^2 \omega/2) \, du^2 + (\sec^2 \omega/2) \, dv^2,$$

where the coordinate lines are geodesic ellipses and hyperbolas. Also show the converse theorem, that if ds^2 can be written in this form, the coordinate lines are geodesic ellipses and hyperbolas.

14. Derive Gauss' form of the differential equation of the geodesics:

$$\sqrt{EG - F^2} \, d\theta = F \, d \ln \sqrt{E} + \tfrac{1}{2}(E_v \, du - G_u \, dv) - F_u \, du,$$

where θ is the angle of the geodesic with the curve $v =$ constant. Compare this formula with Liouville's formula (1–13) and with Exercise 4.

15. *Rectifying developable.* Every space curve is a geodesic on its rectifying developable. This is the explanation of the term *rectifying*. See p. 72.

CHAPTER 5

SOME SPECIAL SUBJECTS

5-1 Envelopes. The method used in Section 2-4 to determine developable surfaces as envelopes of a single infinity of planes can be generalized to singly infinite families of more general surfaces. The method in the case of planes can be summarized (see Chapter 2, Eqs. (4-2) and (4-3)) by stating that if the family of planes is given by

$$F(x, y, z, u) = 0, \tag{1-1}$$

where u is the parameter, and we add to Eq. (1-1) the equations

$$F_u = \partial_u F(x, y, z, u) = 0, \tag{1-2}$$
$$F_{uu} = \partial_{uu}^2 F(x, y, z, u) = 0, \tag{1-3}$$

then, except for special cases, which can be exactly enumerated in this case of a family of planes:

Eqs. (1-1) and (1-2) together for fixed u give the *characteristic line* of a surface (1-1),

Eqs. (1-1), (1-2), (1-3) together for fixed u give the *characteristic point* of the surface (1-1).

Elimination of u from Eqs. (1-1) and (1-2) gives the *envelope* of the family (1-1) as the locus of characteristic lines.

Solving of x, y, z in terms of u from Eqs. (1-1), (1-2), and (1-3) gives the *edge of regression* of the envelope as the locus of characteristic point.

If we now apply this method to the case that Eq. (1-1) represents a general family of surfaces, then we can say in general that *provided this family has an envelope and this envelope has a locus of characteristic points, we can obtain the envelope and the locus by a similar reasoning,* although we may obtain other surfaces, curves, and points as well. Before we pass to the question of proof, let us first take a few examples to elucidate some of the existing possibilities.

(a) The family of spheres of equal radii R with centers on the Y-axis (Fig. 5-1),

$$F \equiv x^2 + (y-u)^2 + z^2 - R^2 = 0, \tag{1-4}$$

has as its envelope the cylinder

$$x^2 + z^2 = R^2$$

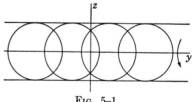

Fig. 5-1

162

obtained by elimination of u from Eq. (1–4) and

$$F_u = -2(y - u) = 0. \tag{1–5}$$

Since

$$F_{uu} = +2,$$

there are no characteristic points.

(b) For the concentric spheres

$$F \equiv x^2 + y^2 + z^2 - u^2 = 0, \tag{1–6}$$

we have

$$F_u = -2u, \qquad F_{uu} = -2.$$

There is no envelope; elimination of u from $F = 0$ and $F_u = 0$ gives the common center.

(c) For the spheres

$$F \equiv (x - u)^2 + y^2 + z^2 - u^2 = x^2 + y^2 + z^2 - 2xu = 0, \tag{1–7}$$

which have their centers on the X-axis and are all tangent to the YOZ-plane (Fig. 5–2), we have

$$F_u = -2x, \qquad F_{uu} = 0.$$

The equations $F = 0$ and $F_u = 0$ give $y^2 + z^2 = 0$, the point of tangency.

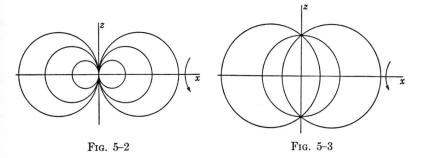

Fig. 5–2　　　　　　　　　　　　Fig. 5–3

(d) For the spheres

$$F \equiv (x - u)^2 + y^2 + z^2 - (a^2 + u^2) = 0, \tag{1–8}$$

which have their centers on the X-axis and all pass through the same circle in the YOZ-plane (Fig. 5–3), we have

$$F_u = -2x, \qquad F_{uu} = 0.$$

The equations $F = 0$ and $F_u = 0$ give the circle through which all spheres pass.

(e) For the cylinders

$$F \equiv (y - u)^2 - x^3 - x^2 = 0,$$

we find

$$F_u = -2(y - u), \qquad F_{uu} = 2.$$

Elimination of u from $F = 0$, $F_u = 0$ gives the envelope $x = -1$, but also the locus $x = 0$ of the nodal lines.

There exists no theory which systematically accounts for all cases which may occur. We have to be satisfied, in the main, with the principle obtained by the generalization of the method used for families of planes. However, it is possible to express the simpler applications of the principle in a more careful way, which we shall now give in outline.

Let us suppose that the surfaces S, given by $F = 0$, have an envelope E. This means that a surface E exists which is tangent to each of the surfaces S along a curve C. We can label the curves C by means of the same parameter u which determines the surface S on which it lies. Then E has the equation $\mathbf{x} = \mathbf{x}(t, u)$, where $\mathbf{x} = \mathbf{x}(t, u_0)$ is the equation of the curve C on the surface S for which $u = u_0$. When the coordinates x, y, z of $\mathbf{x}$ are substituted into $F(x, y, z, u) = 0$ this equation must be identically satisfied.

The vectors $\mathbf{x}_t$ and $\mathbf{x}_u$ determine the tangent plane to E at a point P of C. But the coordinates of these vectors must satisfy the equations

$$F_x \frac{\partial x}{\partial t} + F_y \frac{\partial y}{\partial t} + F_z \frac{\partial z}{\partial t} = 0,$$

$$F_x \frac{\partial x}{\partial u} + F_y \frac{\partial y}{\partial u} + F_z \frac{\partial z}{\partial u} + F_u = 0,$$

or

$$(F_x + F_z p) \frac{\partial x}{\partial t} + (F_y + F_z q) \frac{\partial y}{\partial t} = 0,$$

$$(F_x + F_z p) \frac{\partial x}{\partial u} + (F_y + F_z q) \frac{\partial y}{\partial u} + F_u = 0. \qquad (1-9)$$

Let us now suppose that we take such points of space for which $F = 0$, but $F_u \neq 0$. Then, according to the theory of implicit functions, we can solve $F = 0$ for u and obtain an expression of the form $u = u(x, y, z)$. Since

$$F_x + F_z p = 0, \qquad F_y + F_z q = 0, \qquad (1-10)$$

we can find, by substitution of u into these equations, values for p and q. Suppose that we confine our attention to regions where both $\mathbf{x}_t$, $\mathbf{x}_u$, and p and q are uniquely determined and different from each other, so that the tangent planes to the surfaces S are uniquely determined at each point.

Comparison of Eqs. (1–9) and (1–10) shows that when $F_u \neq 0$ the tangent plane to S cannot be a tangent plane to E. This can only happen when $F_u = 0$. Moreover, if $F_u = 0$ and $\mathbf{x}_t \times \mathbf{x}_u \neq 0$, and if p, q are uniquely determined, then E is actually an envelope of the surfaces S. We have thus found the theorem:

When the equation $F(x, y, z, u) = 0$ represents a family of surfaces S, and the equations $F = 0$, $F_u = 0$ in a certain region $R(x, y, z, u)$ represent for fixed u a family of curves C which form a surface E for varying u, then this surface E is an envelope of S which is tangent to the S along the curves C, provided the surfaces S have in R a unique tangent plane at every point of the curves C.

The curves C are the *characteristics* of the surfaces S.

We can now show that this envelope E actually exists in such regions $R(x, y, z, u)$ in which the vector product $(\text{grad } F) \times (\text{grad } F_u) \neq 0$. In such regions it is possible to solve the equations $F = 0$, $F_u = 0$ for two variables and express them in terms of the two other ones. Let us single out y and z, so that the result appears in the form:

$$y = y(x, u), \qquad z = z(x, u) \qquad \text{(hence here } F_y F_{zu} - F_z F_{yu} \neq 0). \quad (1\text{–}11)$$

Eqs. (1–11) represent a surface which has a tangent plane given at each of its points by the equations

$$\mathbf{i} + \mathbf{j} \frac{\partial y}{\partial x} + \mathbf{k} \frac{\partial z}{\partial x} = \mathbf{i} + \mathbf{j} y_x + \mathbf{k} z_x,$$
$$\mathbf{i} + \mathbf{j} \frac{\partial y}{\partial u} + \mathbf{k} \frac{\partial z}{\partial u} = \mathbf{i} + \mathbf{j} y_u + \mathbf{k} z_u. \quad (1\text{–}12)$$

Since the Eqs. (1–11), substituted into $F = 0$, $F_u = 0$, change these equations into identities, we also have the relations:

$$F_x + F_y y_x + F_z z_x = 0, \qquad F_u + F_y y_u + F_z z_u = 0,$$
$$F_{xu} + F_{yu} y_x + F_{zu} z_x = 0, \qquad F_{uu} + F_{yu} y_u + F_{zu} z_u = 0. \quad (1\text{–}13)$$

These equations determine y_x, z_x and y_u, y_v uniquely, since $F_y F_{zu} - F_z F_{yu} \neq 0$, provided $F_{uu} \neq 0$. The vectors $\mathbf{i} + \mathbf{j} y_x + \mathbf{k} z_x$ and $\mathbf{i} + \mathbf{j} y_u + \mathbf{k} z_u$ are therefore perpendicular to the vector $F_x \mathbf{i} + F_y \mathbf{j} + F_z \mathbf{k}$, which means that the tangent plane to surface (1–11) coincides with the tangent plane to the surface $F = 0$ at all points (x, u) common to both surfaces. For $u = u_0$ we obtain the characteristic on the surface $F(x, y, z, u_0) = 0$. In other words:

In a region $R(x, y, z, u)$ where the vector product $(\text{grad } F) \times (\text{grad } F_u) \neq 0$ and $F_{uu} \neq 0$ there exists an envelope E, locus of the characteristics C.

We can obtain some more information when we consider such regions R_1 of R, where also the triple scalar product $(\text{grad } F, \text{grad } F_u, \text{grad } F_{uu}) \neq 0$. In this case let us solve

$$F = 0, \, F_u = 0, \, F_{uu} = 0$$

for x, y, z in terms of u, which is possible since the functional determinant of F, F_u, F_{uu} is different from zero. The curve thus obtained lies on the surface given by $F = 0$, $F_u = 0$, and hence on the envelope. The tangent direction is given by the vector

$$\mathbf{i}\frac{dx}{du} + \mathbf{j}\frac{dy}{du} + \mathbf{k}\frac{dz}{du} = \mathbf{i}x_u + \mathbf{j}y_u + \mathbf{k}z_u.$$

By a reasoning similar to that used before, we find that

$$F_x x_u + F_y y_u + F_z z_u + F_u = F_x x_u + F_y y_u + F_z z_u = 0,$$
$$F_{xu} x_u + F_{yu} y_u + F_{zu} z_u + F_{uu} = F_{xu} x_u + F_{yu} y_u + F_{zu} z_u = 0,$$

and these equations for $u = u_0$ express the fact that the curve is tangent to the characteristic on the surface $F(x, y, z, u_0) = 0$. Hence:

If there exists an envelope E, locus of characteristics C, then in regions R_1 of R, where the triple scalar product $(\text{grad } F, \text{grad } F_u, \text{grad } F_{uu}) \neq 0$, the equations $F = 0$, $F_u = 0$, $F_{uu} = 0$ together determine a curve which at each of its points is tangent to one of the characteristics C.

This curve is called the *edge of regression* of the envelope; its points are sometimes called the *focal points*.

As an example of an envelope of surfaces which are not plane we can take the *canal surfaces*, already studied by Monge. We obtain them as envelopes of spheres of constant radius a of which the center moves along a space curve $\mathbf{x} = \mathbf{x}(s)$ ($s = u$ is the parameter). Then

$$F \equiv (\mathbf{y} - \mathbf{x}) \cdot (\mathbf{y} - \mathbf{x}) - a^2 = 0.$$

Then, in accordance with Eqs. (1–2) and (1–3) we take

$$F_u = -2\mathbf{t} \cdot (\mathbf{y} - \mathbf{x}) = 0,$$
$$F_{uu} = -2\kappa\mathbf{n} \cdot (\mathbf{y} - \mathbf{x}) + 2 = 0.$$

These equations show that the characteristic is the great circle in which a sphere is intersected by the normal plane to the center curve. The

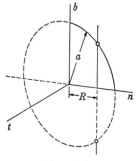

Fig. 5–4

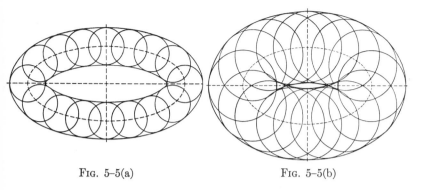

FIG. 5–5(a)   FIG. 5–5(b)

envelope is the locus of these circles and is the canal surface. Since
$F_{uu} = 0$ is equivalent to

$$(\mathbf{y} - \mathbf{x}) \cdot \mathbf{n} = R,$$

we see that the edge of regression is the locus of the points in which the
spheres are intersected by the polar axis of the curve. Hence it splits into
two curves, which come together at points where $a = R$ (Fig. 5–4). When
$a > R$, no edge of regression exists. An example of a canal surface is the
ordinary torus. When the curve $\mathbf{x} = \mathbf{x}(u)$ is a plane curve, the edge of
regression is in the same plane and is the envelope of the circles of constant
radius a with their center on the curve. In Figs. 5–5a, b we find the case
illustrated, where the curve is an ellipse; this figure is the projection on
the XOY-plane of a torus of which the center circle is inclined with respect
to the XOY-plane.

The name *canal surface* is sometimes used for the envelope of any single
infinite system of spheres. One of the center surfaces (Section 2–9) of
these surfaces degenerates into a curve.

EXERCISES

1. A necessary condition for a family of curves $F(x, y, u) = 0$ to have an envelope is that the equations $F = 0$, $F_u = 0$ have a common solution. If in a certain neighborhood $(x_0, y_0, u_0)F_xF_{yu} - F_yF_{xu} \neq 0$, then we obtain the envelope of
the curves $F = 0$, $F_u = 0$ by the elimination of u from $F = 0$, $F_u = 0$.

2. Determine the envelope of the following plane curves: (a) the circles of
constant radius, of which the centers are on a line, (b) the circles $(x - a)^2 + y^2 = b^2$,
$b^2 = 4am$ (m constant).

3. *A problem in elementary ballistics.* A projectile is fired from a gun with
constant velocity v_0 making an angle α with the horizon. Show that the envelope
of the trajectories, if α varies in a vertical plane, is a parabola. This problem was

taken up and solved by E. Torricelli in his book on the motion of projectiles (1644).

4. From the points of a parabola $y^2 = 4x - 1$ tangents are drawn to the ellipse $\dfrac{x^2}{a^2} + \dfrac{y^2}{b^2} = 1$. Find the envelope of the chords connecting the points of contact. This envelope is called the *polar figure* of the parabola with respect to the ellipse.

5. Show that a plane curve is the envelope of its osculating circles. This is a case where, for the envelope, the equations $F = 0$, $F_u = 0$ hold, although two osculating circles do not intersect along an arc of increasing (decreasing) κ.

6. Show that a space curve is the edge of regression on the canal surface formed by its osculating spheres.

7. Verify that the edge of regression of a canal surface is tangent to the characteristic.

8. Analyze the cases (a) $-$ (e) of this chapter by evaluating $(\text{grad } F) \times (\text{grad } F_u)$ and $(\text{grad } F,\, \text{grad } F_u,\, \text{grad } F_{uu})$.

9. Find the envelope of the family of spheres which have as their diameters a set of parallel chords of a circle.

10. When the envelope S of a family of surfaces is intersected by a surface S_1 which is not tangent to the edge of regression E, then the curve of intersection C of S and S_1 has a cusp (or higher singularity) at the point where S intersects E. Hint: Show that at that point $\mathbf{x}' = 0$, when C is given by $\mathbf{x} = \mathbf{x}(u)$.

11. When a family of surfaces is given by the equation $\mathbf{x} = \mathbf{x}(u, v, \alpha)$, where α is a parameter and u, v are curvilinear coordinates, then the envelope, if it exists, is found by the elimination of α from $\mathbf{x} = \mathbf{x}(u, v, \alpha)$ and $(\mathbf{x}_u\, \mathbf{x}_v\, \mathbf{x}_\alpha) = 0$. See, on the discussion of envelopes from this point of view, O. Bierbaum, *Festschrift der Techn. Hochschule Brünn*, 1899, pp. 117–150.

5-2 Conformal mapping. Two surfaces S and S_1 are said to be *mapped* upon one another if there exists a one-to-one correspondence between their points. When S is given by the equation $\mathbf{x} = \mathbf{x}(u, v)$ and S_1 by the equation $\mathbf{x}_1 = \mathbf{x}_1(u_1, v_1)$, then a mapping of a region of S on a corresponding region of S_1 is established by the relations

$$u_1 = u_1(u, v), \qquad v_1 = v_1(u, v), \tag{2-1}$$

where u_1, v_1 are single-valued functions with continuous partial derivatives such that the functional determinant $\partial(u_1, v_1)/\partial(u, v) \neq 0$. When we use Eq. (2–1) to transform the coordinates (u_1, v_1) of S_1 into coordinates (u, v) on the same surface S_1, then we obtain the equation of S_1 in the form $\mathbf{x}_1 = \mathbf{x}_1(u, v)$. Two points of S and S_1 which correspond in the mapping are now indicated by the same values of (u, v):

$$\mathbf{x} = \mathbf{x}(u, v); \qquad \mathbf{x}_1 = \mathbf{x}_1(u, v). \tag{2-2}$$

We shall discuss the following types of mapping:

I. Conformal mapping, in which angles are preserved;

II. Isometric mapping, in which distances and angles are preserved;

III. Equiareal mapping, in which areas are preserved;

IV. Geodesic mapping, in which geodesics are preserved.

Conformal mapping. A mapping of surface S on surface S_1 (we shall use this expression although we mean the mapping of a certain region of S on a corresponding region of S_1) is *conformal* if the angle between two directed curves through a point P of S is equal to the angle between two corresponding directed curves through the point P_1 on S_1 corresponding to P. When the two surfaces are written in the form (2–2) and all elements of S_1 corresponding to those on S are indicated by the index 1, then for the angles θ, θ_1 between the corresponding elements $d\mathbf{x}$, $\delta\mathbf{x}$, $d_1\mathbf{x}$, $\delta_1\mathbf{x}$ the equation holds: $\theta = \theta_1$. If we take as one pair of corresponding elements $\delta\mathbf{x}$, $\delta_1\mathbf{x}$, the elements (δu arbitrary, $\delta v = 0$), we find that according to Chapter 2, Eq. (2–9):

$$\frac{E \, du \, \delta u + F \, dv \, \delta u}{ds\sqrt{E \, \delta u^2}} = \frac{E_1 \, du \, \delta u + F_1 \, dv \, \delta u}{ds_1\sqrt{E_1 \, \delta u^2}},$$

or

$$\sqrt{E}\frac{du}{ds} + \frac{F}{\sqrt{E}}\frac{dv}{ds} = \sqrt{E_1}\frac{du}{ds_1} + \frac{F_1}{\sqrt{E_1}}\frac{dv}{ds_1}.$$

This equation must hold for all directions dv/du, hence

$$\sqrt{E}\frac{du}{ds} = \sqrt{E_1}\frac{du}{ds_1}, \qquad F\frac{dv}{\sqrt{E}\,ds} = F_1\frac{dv}{\sqrt{E_1}\,ds_1},$$

or

$$\frac{ds_1}{ds} = \frac{\sqrt{E_1}}{\sqrt{E}} = \frac{F_1\sqrt{E}}{F\sqrt{E_1}}.$$

This equation shows that the ratio $\rho = ds_1/ds$ is independent of dv/du and depends only on u and v. Hence

$$ds_1 = \rho \, ds, \qquad \rho = \rho(u, v), \tag{2–3}$$

or

$$E_1 = \rho^2 E, \qquad F_1 = \rho^2 F, \qquad G_1 = \rho^2 G. \tag{2–4}$$

Conversely, when Eq. (2–3) holds, then because of Eq. (2–4):

$$\cos \theta_1 = \frac{(E \, du \, \delta u + F(du \, \delta v + dv \, \delta u) + G \, dv \, \delta v)}{ds \, \delta s} = \cos \theta.$$

We have thus arrived at the result:

A necessary and sufficient condition that a mapping be conformal is that the elements of arc ds, δs at corresponding points be proportional.

We can also express this property by stating that a small figure (e.g., a triangle) at a point of a surface is almost similar to the figure which corresponds to it in the mapping, and the more similar the smaller the figures are:

A necessary and sufficient condition that a mapping be conformal is that it be a similarity "in the infinitesimal."

When ρ is constant the ratio of similarity is the same for all points. We then speak of *similitude.*

A conformal map of a small region of a surface near a point on a plane is therefore very nearly accurate in the angles as well as in the ratio of distances, although the map may give a very distorted picture of the region in the large.

From Eq. (2–3) it follows that from $ds = 0$ follows $ds_1 = 0$. If, therefore, we admit imaginary elements, then we find that in a conformal mapping isotropic elements (Section 1–12) correspond. Discarding isotropic developables ($EG - F^2 = 0$, see Section 5-6) we have on each surface a net of isotropic curves which we can take as the net of coordinate lines. Then $E = G = 0$ and

$$ds^2 = 2F\,du\,dv, \qquad ds_1^2 = 2F_1\,du\,dv, \qquad F_1 = \rho^2 F. \qquad (2-5)$$

This equation shows that when the isotropic curves correspond, ds and ds_1 are proportional:

A necessary and sufficient condition that a mapping be conformal is that the isotropic curves correspond.

Since this correspondence can be accomplished for any pair of surfaces, we find that *two surfaces can always be mapped conformally upon each other.*

This has to be understood again in the sense that to a certain region on surface S there corresponds a certain region on S_1 on which it can be conformally mapped. The theory of complex variables establishes exact conditions for the character of these regions and their boundaries. The fundamental theorem of Riemann states that any region with a suitable boundary can be conformally mapped on a circle by means of a simple analytic function.*

After having introduced the isotropic curves, let us again change the coordinate curves on both surfaces by the transformation

$$u = u_1 + iv_1, \qquad v = u_1 - iv_1. \qquad (2-6)$$

---

* See, e.g., E. C. Titchmarsh, *The theory of functions*, Oxford, Clarendon Press, 1932, p 207.

The first fundamental forms in the new coordinates, written in terms of the new coordinates, are now

$$ds^2 = 2F(du_1^2 + dv_1^2), \qquad ds_1^2 = 2F_1(du_1^2 + dv_1^2), \qquad (2\text{–}7)$$

where F and F_1 are now functions of u_1 and v_1.

If a system of curves can be introduced as coordinate curves so that ds^2 takes the form

$$ds^2 = \lambda(du^2 + dv^2), \qquad \lambda = \lambda(u, v), \qquad (2\text{–}8)$$

then this system is called an *isometric* (or *isothermic*) *system*. It is necessarily orthogonal. Since the length of du is the same as the length of dv, we can sketch the nature of isometric systems by saying that they *divide the surface into infinitesimal squares*. Eq. (2–7) expresses the fact that two surfaces are conformally mapped upon each other when an isometric system on one surface corresponds to an isometric system on the other surface.

> The use of the term "isometric" both for certain systems of curves and for a certain type of mapping does not mean that there is any direct relation between them.

We shall now show how on a surface an infinity of isometric systems can be obtained. For this purpose, let us suppose that apart from the system obtained by Eqs. (2–5) and (2–6) and expressed by Eq. (2–8) there exists another isometric system, given in terms of (u, v) by

$$\alpha = \alpha(u, v) = \text{constant}, \qquad \beta = \beta(u, v) = \text{constant}. \qquad (2\text{–}9)$$

Then there exist two functions $\lambda(u, v)$, $\Lambda(\alpha, \beta)$ such that

$$ds^2 = \lambda(du^2 + dv^2) = \Lambda(d\alpha^2 + d\beta^2).$$

But $(\alpha_u = \partial\alpha/\partial u, \text{ etc.})$:

$$d\alpha^2 + d\beta^2 = (\alpha_u^2 + \beta_u^2)\, du^2 + 2(\alpha_u\alpha_v + \beta_u\beta_v)\, du\, dv + (\alpha_v^2 + \beta_v^2)\, dv^2.$$

The necessary and sufficient conditions that the system (2–9) be isometric are therefore:

$$\alpha_u^2 + \beta_u^2 = \alpha_v^2 + \beta_v^2; \qquad \alpha_u\alpha_v + \beta_u\beta_v = 0.$$

If we write, for a moment, $\mu = \alpha_u/\beta_u = -\beta_v/\alpha_v$, then

$$(\mu^2 + 1)\alpha_v^2 = (\mu^2 + 1)\beta_u^2.$$

Now $\mu^2 + 1 \neq 0$, since $\mu = \pm i$ leads to a linear relation between α and β. Hence either

$$\alpha_v = \beta_u, \qquad \alpha_u = -\beta_v, \qquad (2\text{–}10)$$

or

$$\alpha_v = -\beta_u, \qquad \alpha_u = \beta_v. \tag{2-11}$$

Eqs. (2–10) and (2–11) are known as the *Cauchy-Riemann equations*, and are the conditions that, in the case (2–11) $\alpha + i\beta$, and in the case (2–10) $\alpha - i\beta$, is an analytic function of the complex variable $u + iv$:

$$\alpha \pm i\beta = f(u + iv).^*$$

This leads us to the theorem:

When $u = $ constant, $v = $ constant form a system of isometric coordinates, then all other isometric systems are given by

$$Rf(u + iv) = \alpha, \qquad If(u + iv) = \beta,$$

where $Rf(u + iv)$ and $If(u + iv)$ are the real and imaginary parts of an arbitrary analytic function of the complex variable $u + iv$.

If, for example, we take $\alpha + i\beta = (u + iv)^3$, then the system defined by $u^3 - 3uv^2 = $ constant, $3u^2v - v^3 = $ constant forms an isometric system, if $u = $ constant, $v = $ constant forms such a system.

With the aid of these isometric systems we can map one surface upon another in an infinity of ways, depending on an arbitrary analytic function of a complex variable. If, namely, we refer a surface S to an isometric system of coordinates (u, v), and another surface likewise to an isometric system of coordinates (u_1, v_1), then the correspondence $u = u_1$, $v = v_1$ establishes a conformal mapping of S upon S_1.

Rectangular cartesian coordinates (x, y) in the plane form an isometric system. Consequently, *any mapping of a surface upon the plane such that an isometric system of the surface corresponds to the lines $x = $ constant, $y = $ constant in the plane is a conformal mapping, and these mappings are the only conformal mappings of the surface on the plane.* Any analytic function of a complex variable performs such a mapping.

EXAMPLE. For a rotation surface (Chapter 2, Eq. (2–13)):

$$ds^2 = (1 + f'^2)\, dr^2 + r^2\, d\varphi^2 = r^2 \left(\frac{1 + f'^2}{r^2}\, dr^2 + d\varphi^2 \right).$$

This ds^2 can be transformed to isometric coordinates (u, v) by the transformation:

$$u = \int \frac{\sqrt{1 + f'^2}}{r}\, dr, \qquad v = \varphi.$$

---

* See, e.g., C. E. Titchmarsh, *The theory of functions*, Oxford, Clarendon Press, 1932, p. 68.

A rotation surface can therefore be conformally mapped upon the plane by the transformation

$$x = \varphi, \qquad y = u,$$

which transforms the meridians φ = constant into lines parallel to the Y-axis, and the parallels u = constant into lines parallel to the X-axis. Equally spaced meridians pass into equally spaced lines x = constant, but the spacing of the parallels is changed.

In particular, if $f(r) = \sqrt{a^2 - r^2}$, $r = a \cos \theta$, $f(r) = a \sin \theta$, we obtain in

$$x = \varphi, \qquad y = \int \sec \theta \, d\theta = \ln \tan\left(\frac{\theta}{2} + \frac{\pi}{4}\right)$$

a conformal representation of the sphere on the plane, in which $\varphi = 0$, $\theta = 0$ corresponds to $x = 0$, $y = 0$; the equator $\theta = 0$ to the X-axis; and equally spaced parallels θ = constant correspond to lines parallel to the X-axis at ever-increasing distance when θ increases, until North and South pole are mapped at infinity. This is the *Mercator projection* (Fig. 5–6), which therefore, in the way in which it is commonly used in our atlases for maps of the world, is quite faithful near the equator, but gives an exaggerated impression of the dimensions of Arctic regions. Loxodromes on the sphere are mapped into straight lines on the map (Section 2–2).

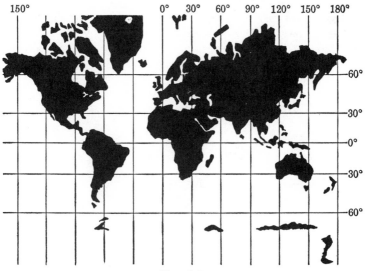

Fig. 5–6

Gerhard Mercator, latinized for *Kremer* (1512–1594), belongs with such men as Ortelius and Blaeu to the school of great Flemish-Dutch cartographers of the sixteenth and seventeenth centuries. He introduced the "Mercator" projection in his famous world map of 1569 (on which the term "Norumbega" is used for a section of present New England). This projection had already occasionally been used. Mercator was aware of the conformal character of his projection.

Another conformal map of the sphere on the plane is found by transforming by means of

$$u = \ln r_1, \qquad \varphi = \varphi_1$$

the system of polar coordinates of the plane (r_1, φ_1) into an isometric system:

$$ds^2 = dr_1^2 + r_1^2 \, d\varphi_1^2 = r_1^2(du^2 + d\varphi_1^2) = e^{2u}(du^2 + d\varphi_1^2).$$

The resulting conformal mapping is accomplished by

$$u = \int \sec \theta \, d\theta = \ln \tan\left(\frac{\theta}{2} + \frac{\pi}{4}\right) + c, \qquad \varphi_1 = \varphi,$$

or, taking the integrating constant $c = \ln 2a$:

$$r_1 = 2a \tan\left(\frac{\theta}{2} + \frac{\pi}{4}\right), \qquad \varphi_1 = \varphi.$$

This is the *stereographic projection,* and can be obtained by projecting the points of the sphere from the North pole on the tangent plane at the

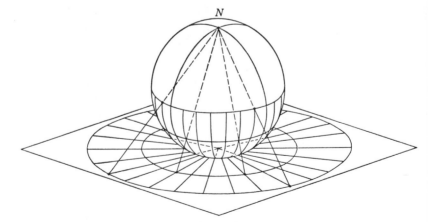

Fig. 5–7

FIG. 5-8

South pole (Fig. 5–7). Meridians pass into straight lines through the South pole S, parallels into concentric circles with S as center. The spacing of the meridians is preserved, but equally spaced parallels are mapped into circles at ever-increasing distance when θ decreases from $\pi/2$ to $-\pi/2$. The projection is quite faithful near the South pole; the North pole is at infinity. Loxodromes on the sphere are mapped into logarithmic spirals. Stereographic projection from S to the tangent plane at N is obtained by changing θ into $-\theta$ (Fig. 5–8).

This projection was known to Ptolemy (c. 150 A.D.), who described it in his *Geographia;* it may be due to Hipparch (c. 150 B.C.). It was used for map projections by the Flemish mathematician Gemma Frisius (1540). The name was introduced in a book on optics by the Belgian author F. d'Aiguillon (1613).

Our theory of conformal mapping is due to Gauss, whose paper on the representation "of a given surface upon another such that the map is similar in its smallest parts" appeared in 1822 (*Werke* IV, pp. 193–216; partial English translation in D. E. Smith, *Source book in mathematics*, 1922, pp. 463–475). The name *isothermic* is due to G. Lamé (1833); see his *Leçons sur les coordonnées curvilignes* (1859). The name *isometric* is due to Bonnet.

5-3 Isometric and geodesic mapping. A mapping (2–2) is *isometric* when at two correspondent points the first differential forms are the same. Hence $E = E_1$, $F = F_1$, $G = G_1$, or

$$E\,du^2 + 2F\,du\,dv + G\,dv^2 = E_1\,du^2 + 2F_1\,du\,dv + G_1\,dv^2. \qquad (3\text{–}1)$$

We have already met such mappings when we discussed bending (Section 3–3) and surfaces of constant curvature (Section 4–5). In Section 3–4 we mapped a catenoid isometrically on a right helicoid.

The fact that isometry involves the preservation of certain invariants such as the Gaussian curvature implies that two arbitrary surfaces cannot, as a rule, be isometrically mapped upon each other. There are *classes of isometric surfaces*, such as developable surfaces or surfaces of Gaussian curvature 1. Surfaces which can be transformed into each other by bending are isometric, and in this case the surfaces are also called *applicable*. The terms applicable and isometric are often identified, but it is not *a priori* clear that the existence of an isometric correspondence between two surfaces also means that one surface can pass into the other by bending, that is, by a continuous isometric transformation. The relations between applicability and isometry therefore need careful investigation. Results on the local problem have been reached in the case of surfaces of zero curvature, where we know that the two concepts are identical (Section 4–6). A much larger class of surfaces is covered by a theorem of E. E. Levi, which states that two isometric surfaces of zero or negative curvature are always applicable provided certain conditions of differentiability are satisfied. Two isometric surfaces of positive curvature either are applicable to each other or one surface is applicable to a surface symmetrical to the other; in this case the surfaces must be analytic.

Levi's paper, written at the request of L. Bianchi, can be found *Atti Accad. Torino* **43,** 1907–08, pp. 292–302.

The problem of isometric mapping can be conceived in two different ways:

1. *Given two surfaces S and S_1 with given first fundamental forms*

$$ds^2 = E\,du^2 + 2F\,du\,dv + G\,dv^2, \qquad ds_1^2 = E_1\,du_1^2 + 2F\,du_1\,dv_1 + G_1\,dv_1^2,$$

to find whether there exists a correspondence

$$u_1 = u_1(u, v), \qquad v_1 = v_1(u, v), \tag{3–2}$$

such that $ds^2 = ds_1^2$.

2. *Given a positive definite quadratic form $E\,du^2 + 2F\,du\,dv + G\,dv^2$, to find all surfaces for which this form can be considered as the first differential form.* In other words, assuming that there is a surface S for which the given form is the ds^2, to find all surfaces which can be isometrically mapped on S.

The first problem is sometimes called after *Minding*. It is the simpler of the two problems, and can be solved in each particular case by means of a finite sequence of differentiations and eliminations. We have solved it in Chapter 4 for the case that S and S_1 have constant Gaussian curvature, where we found ∞^3 transformations (3–2) which establish the correspondence. However, we cannot conclude from this case, in which equality of Gaussian curvature means isometry, that two surfaces can always be isometrically mapped upon each other if a correspondence (3–2) can be found such that the Gaussian curvature at corresponding points is equal. For instance, the surfaces

$$
\begin{aligned}
x &= u\cos v, & y &= u\sin v, & z &= \ln u & \text{(a rotation surface)}\\
x &= u\cos v, & y &= u\sin v, & z &= v & \text{(a right helicoid)}
\end{aligned}
\tag{3–3}
$$

have at corresponding points (u, v) the same Gaussian curvature $K = -(1 + u^2)^{-2}$, but are not isometric.

Criteria for the possibility of the isometric mapping of two surfaces were first established by F. Minding, *Journal für Mathem.* **19**, 1839, pp. 370–387. For an exposition of the theory, see L. P. Eisenhart, *Differential geometry*, pp. 323–325.

The second problem is sometimes called after *Bour*. We have solved it for the case that the differential form is of zero curvature, e.g., $du^2 + dv^2$, when the answer is given by the totality of developable surfaces. When we ask, in this case, for all solutions of the form $z = f(x, y)$, then the answer is given by all solutions of the differential equation $rt - s^2 = 0$ (Exercise 2, Section 2–8). In the case of differential forms of constant curvature we had to be satisfied with a sketch of those solutions which are surfaces of rotation (Section 4–6). The great variety of solutions in this particular case may give some idea of the complexity of Bour's problem. It leads, in the general case, to the study of the solutions of a partial differential equation of the Monge-Ampère type:

$$rt - s^2 + Ar + Bs + Ct + D = 0, \qquad (r = \partial^2 z/\partial x^2, \text{ etc.})$$

where A, B, C, D are functions of x, y, z, p and q.* In the case that the coordinate lines are isotropic:

$$ds^2 = 2F\, du\, dv, \qquad F = F(u, v),$$

this equation is $(p = \partial x/\partial u, \text{ etc.})$:

$$rt - s^2 - qr(\partial_v \ln F) - pt(\partial_u \ln F)$$
$$= (\partial^2_{uv} \ln F)\, (F - 2pq) - pq(\partial_u \ln F)(\partial_v \ln F).$$

The solutions of this equation have only been fully investigated in some special cases, such as paraboloids of revolution. For $F = $ constant we obtain the developable surfaces.

This problem received attention when in 1859 the Paris Academy proposed it as a subject of a contest. The prize went to Edouard Bour (1832–1866), whose paper, according to Liouville, "could be taken as a beautiful memoir by Lagrange." The principal part of this paper was published in the *Journal Ecole Polytechn.* **22** (1862), pp. 1–148. Honorably mentioned were papers by Bonnet and Codazzi. For an exposition of Bour's problem see Darboux' *Leçons III*, also L. P. Eisenhart, *Differential geometry*, Chap. 9.

Geodesic mapping preserves geodesics. An isometric mapping is also geodesic, but not every geodesic mapping is isometric. Since two arbitrary surfaces cannot in general be geodesically mapped upon each other,

---

* In E. Goursat, *Leçons sur l'intégration des équations aux dérivées partielles du second ordre* (1896), p. 39, such equations of the Monge-Ampère type are investigated.

we can ask for such classes of isometric surfaces which can be mapped geodesically on other classes of isometric surfaces. This problem has been fully solved. We first prove a special case of this theorem.

The only surfaces which can be geodesically mapped upon the plane are those of constant curvature. This theorem, discovered by Beltrami, is proved by referring the surface to that system of coordinates (u, v) which in the mapping corresponds to cartesian coordinates in the plane. The geodesics of the surface are then linear expressions in u and v, and conversely, if there exists a coordinate system (u, v) on the surface in which the geodesics are expressed in the form $au + bv + c = 0$ (a, b, c constants), then it can be geodesically mapped on the plane. This means that the equation of the geodesic lines (Chapter 4, Eq. (2–3a)) must be identically satisfied when $v'' = d^2v/du^2 = 0$:

$$\Gamma^2_{11} = \Gamma^1_{22} = 0, \qquad \Gamma^1_{11} = 2\Gamma^2_{12}, \qquad \Gamma^2_{22} = 2\Gamma^1_{12}.$$

The Gaussian curvature K then satisfies the equations (see Chapter 3, Eqs. (3–3)):

$$\begin{aligned}
KE &= \Gamma^2_{12}\Gamma^2_{12} - \partial_u\Gamma^2_{12}, & KF &= \Gamma^1_{12}\Gamma^2_{12} - \partial_v\Gamma^2_{12}, \\
KG &= \Gamma^1_{12}\Gamma^1_{12} - \partial_v\Gamma^1_{12}, & KF &= \Gamma^1_{12}\Gamma^2_{12} - \partial_u\Gamma^1_{12}.
\end{aligned} \qquad (3\text{--}4)$$

Since $\partial^2_{vu}\Gamma^2_{12} = \partial^2_{uv}\Gamma^2_{12}$, we obtain by differentiating the first two equations of (3–4) and using (3–4) again:

$$\begin{aligned}
EK_v - FK_u =& - K(E_v - F_u) + 2\Gamma^2_{12}\partial_v\Gamma^2_{12} - \Gamma^1_{12}\partial_u\Gamma^2_{12} - \Gamma^2_{12}\partial_u\Gamma^1_{12} \\
=& -K(E_v - F_u) + 2\Gamma^2_{12}\Gamma^1_{12}\Gamma^2_{12} - 2KF\Gamma^2_{12} - \Gamma^1_{12}\Gamma^2_{12}\Gamma^2_{12} \\
& \qquad + KE\Gamma^1_{12} - \Gamma^2_{12}\Gamma^1_{12}\Gamma^2_{12} + KF\Gamma^2_{12} \\
=& -K(E_v - F_u) + K(-F\Gamma^2_{12} + E\Gamma^1_{12}).
\end{aligned}$$

Substituting into this equation the expressions for the Christoffel symbols, Chapter 3, Eqs. (2–7), we finally conclude (see Exercise 20, Section 3–3):

$$EK_v - FK_u = -K(E_v - F_u) + K(E_v - F_u) = 0.$$

Differentiating the last two equations of Eq. (3–4), we obtain in a similar way:

$$FK_v - GK_u = 0.$$

Hence

$$K_u = K_v = 0,$$

which shows that K is constant.

Conversely, if K is constant, then geodesic mapping is possible. This can be shown by mapping the sphere geodesically on the plane by means of the relations

$$x = \cot \theta \cos \varphi, \qquad y = \cot \theta \sin \varphi, \tag{3-5}$$

where θ and φ are, as usual, the latitude and longitude of the sphere and (x, y) are rectangular cartesian coordinates in the plane. Indeed, all geodesics of the sphere lie in planes through the center, and are therefore (compare with Chapter 2, Eq. (1–7)) given by:

$$A \cos \theta \cos \varphi + B \cos \theta \sin \varphi + C \sin \theta = 0, \quad (A, B, C, \text{arbitrary constants}),$$

which by virtue of Eq. (3–5) passes into the equation of all straight lines in the plane:

$$Ax + By + C = 0.$$

This transformation maps great circles through a point on the sphere into straight lines through a point in the plane, the meridians $\varphi = $ constant passing into the lines $y = x \tan \varphi$, the parallels $\theta = $ constant into the circles $x^2 + y^2 = \cot^2 \theta$. The plane corresponds to one of the hemispheres $0 < \theta \leqslant \pi/2$ or $-\pi/2 \leqslant \theta < 0$.

As to the problem of the geodesic mapping of arbitrary surfaces, it can be shown that in general the only geodesic mappings are isometric mappings with or without similarity. Excepted are two classes of surfaces, which are specified by the following *theorem of Dini-Lie.*

Two surfaces can be geodesically mapped upon each other without isometry or similarity when their first fundamental forms, by means of corresponding coordinates (u, v), can be cast either into the form of Dini:

$$ds^2 = (U - V)(du^2 + dv^2), \qquad ds_1^2 = \left(\frac{1}{V} - \frac{1}{U}\right)\left(\frac{du^2}{U} + \frac{dv^2}{V}\right), \tag{3-6}$$

or into the form of Lie:

$$ds^2 = (u + V) \, du \, dv, \qquad ds_1^2 = \frac{u + V}{2v^3} \, du \, dv - \frac{(u + V)^2}{4v^4} \, dv^2, \tag{3-7}$$

where U is a function of u only and V a function of v only.

The form (3–6) characterizes the *Liouville surfaces* (see Section 4-2, Exercises 18 and 19). The form (3–7) characterizes certain surfaces discussed by Lie. They are real only for some values of V, and the correspondence itself is imaginary. The only real surfaces which can be geodesically mapped on each other by means of a real transformation without isometry or similarity are given by Eq. (3–6).

For the proof of the theorem of Dini-Lie we refer to A. R. Forsyth, *Lectures on differential geometry*, pp. 243–254, or to Darboux' *Leçons III*, pp. 40–65. Beltrami's theorem was first published in 1865 (*Opere I*, pp. 262–280). It was followed by the paper of U. Dini, *Annali di Matem*, **3** (1869), pp. 269–293, after which S. Lie gave his supplementary theorem in 1879, see *Math. Annalen* **20** (1882), p. 419. (See also Exercise 11 below.)

EXERCISES

1. *Theorem of Tissot.* In any real mapping of real surfaces which is not conformal there exists a unique orthogonal system of curves on each surface to which corresponds an orthogonal system on the other surface (Tissot, *Nouvelles Annales de Mathém.* **17**, 1878, p. 151).

2. A mapping which is both geodesic and conformal is either isometric or a similitude.

3. *Inversion.* If for the two points $P(\mathbf{x})$ and $P_1(\mathbf{x}_1)$ of space the relation $\mathbf{x}_1 = a^2\mathbf{x}/(\mathbf{x} \cdot \mathbf{x})$ exists, we say that we have established an *inversion*. Show that (a) under inversion spheres remain spheres and (b) the mapping established between two surfaces by inversion is conformal.

4. Show that any mapping of one surface upon another which preserves the asymptotic lines also preserves conjugate systems.

5. The sum of the angles of a triangle on a surface of revolution, of which the sides are loxodromes, is equal to two right angles.

6. Two surfaces of constant curvature K_1, K_2, $K_1 \neq K_2$, admit a similitude.

7. Establish a conformal mapping of the pseudosphere of curvature -1, for which (Section 4–6):

$$ds^2 = du^2 + e^{2u}\,dv^2,$$

upon the plane such that the geodesics pass into the circles $x^2 + (y - a)^2 = b^2$.

8. *Equiareal map.* The mapping of two surfaces $\mathbf{x} = \mathbf{x}(u, v)$ and $\mathbf{x}_1 = \mathbf{x}_1(u, v)$ preserves the areas of corresponding figures if and only if

$$EG - F^2 = E_1G_1 - F_1^2.$$

9. Show that if we project every point P of a sphere on the cylinder which is tangent to the sphere at the equator, and then develop the cylinder into a plane, we obtain an equiareal map of the sphere upon the plane. What is the image of the meridians and of the parallels of the sphere?

10. Show that the mapping of the sphere on the plane

$$x = a \sin \theta + f(\varphi), \qquad y = a\phi,$$

where $f(\varphi)$ is arbitrary, is also equiareal.

11. Prove the theorem of Dini (hence the part of the theorem of Section 5–3 expressed by Eq. (3–6)) by selecting on the geodesically mapped surfaces a coordinate system according to Tissot's theorem (Exercise 1), so that $F = F_1 = 0$. Hint: Show that $(E_1/G_1^2):(E/G^2)$ is independent of u, and that $(G_1/E_1^2):(G/E^2)$ is independent of v.

12. Find the complex function by which we obtain the stereographic projection from the Mercator projection.

13. *Polyconic projections.* These are mappings of the sphere upon the plane in which the parallels are mapped into a system of circles whose centers C lie on a straight line corresponding to a meridian (the "central" meridian). In the so-called American polyconic projection:

(a) each parallel is a segment of a circle, the developed base of the cone tangent to the sphere along this parallel, the equator being a straight line,

(b) parallels equally spaced along the central meridian are equally spaced along this meridian in the map,

(c) the scale along the parallels remains constant, so that for all parallel lines the unit of measure is represented by the same distance.

Show that (a) $\alpha = \varphi \sin \theta$, when α is the angle with the central meridian of the line connecting a point P of longitude φ from the central meridian and latitude θ with its corresponding center C,

(b) The projection can be given by

$$x = a \cot \theta \sin \alpha,$$
$$y = a\theta + a \cot \theta (1 - \cos \alpha),$$

(c) The projection is neither conformal nor equiareal.

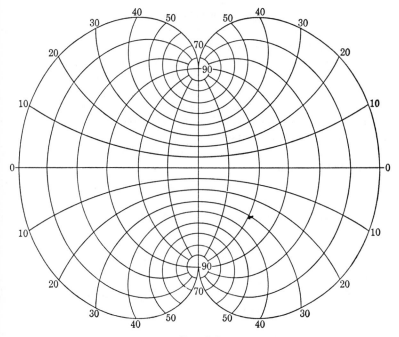

Fig. 5-9

Fig. 5–9 shows a map of the earth in this projection. This projection, extensively used by the U. S. Coast and Geodetic Survey, seems to have been devised by Superintendent F. R. Hassler (1770–1843) for the charting of the coast of the U. S. A. See O. S. Adams, *General theory of polyconic projections*, Department of Commerce, U. S. Coast and Geodetic Survey, Serial No. 110 (1919), especially pp. 143–152.

5–4 Minimal surfaces. We have defined minimal surfaces as surfaces on which the asymptotic lines form an orthogonal system (Section 2–8). This means that the Dupin indicatrix consists of two conjugate rectangular hyperbolas (that is, hyperbolas with perpendicular asymptotes: $x^2 - y^2 = \pm 1$) and that consequently $\kappa_1 = -\kappa_2$, and the mean curvature M is zero (Chapter 2, Eq. (7–2)): *Minimal surfaces are surfaces of zero mean curvature.* The orthogonality of the asymptotic lines can be expressed by the relation (see Chapter 2, Eq. (7–2) or Exercise 6, Section 2–3):

$$Eg - 2Ff + Ge = 0. \tag{4–1}$$

Examples of minimal surfaces have been found in the plane, the catenoid, and the right helicoid (Section 2–8).

Minimal surfaces are sometimes defined as surfaces of the smallest area spanned by a given closed space curve. We have seen in the case of geodesics that this definition by means of a minimum condition of the calculus of variations is not always satisfactory. Consequently we preferred for geodesics the definition as lines of vanishing geodesic curvature. We shall now show that our definition of minimal surfaces as surfaces of zero mean curvature has exactly the same relation to the problem of the surface of minimal area in a given contour C as our definition of geodesics has to the problem of the shortest distance.

To show this, let a very small deformation of a surface $\mathbf{x} = \mathbf{x}(u, v)$ be given by the equation

$$\mathbf{x}_1 = \mathbf{x} + \epsilon\mathbf{N}, \tag{4–2}$$

where ϵ is a small quantity and like $\mathbf{x}$ and $\mathbf{N}$ a function of u and v. Then

$$\partial_u\mathbf{x}_1 = \mathbf{x}_u + \epsilon\mathbf{N}_u + \epsilon_u\mathbf{N},$$
$$\partial_v\mathbf{x}_1 = \mathbf{x}_v + \epsilon\mathbf{N}_v + \epsilon_v\mathbf{N},$$

and we find for the coefficients of the first fundamental form of the deformed surface, neglecting terms of higher order in ϵ:

$$E_1 = E - 2\epsilon e, \qquad F_1 = F - 2\epsilon f, \qquad G_1 = G - 2\epsilon g. \tag{4–3}$$

Hence, introducing the mean curvature M (Chapter 2, Eq. (7–2)):

$$E_1 G_1 - F_1^2 = (EG - F^2) - 2\epsilon(Eg - 2Ff + Ge) = (EG - F^2)(1 - 4\epsilon M),$$

and integrating over an area enclosed by a fixed contour C:

$$\iint \sqrt{E_1 G_1 - F_1^2}\, du\, dv = \iint \sqrt{EG - F^2}\, du\, dv - 2 \iint \epsilon M \sqrt{EG - F^2}\, du\, dv.$$

This formula can be written in the form

$$\delta A = -2 \iint \epsilon M\, dA, \tag{4–4}$$

where dA is the element of area of the original surface and δA is the so-called (first) variation of the area enclosed by the fixed contour C. This formula (4–4) may be compared to that for the (first) variation of the length of a curve (Chapter 4, Eq. (4–1)) fixed at the ends. Where in previous sections we had occasion to compare the ordinary or geodesic curvature of a curve with the Gaussian curvature of a surface, we have here an analogy with the mean curvature.

The first variation δA vanishes for all ϵ if $M = 0$, that is, if the mean curvature vanishes. This can be expressed in the following words:

If there is a surface of minimum area passing through a closed space curve, it is a minimal surface.

We can also find this result (compare with Section 4–4) by means of the general rules of the calculus of variations. Let $z = f(x, y)$ represent a surface; then our problem is to find the Euler-Lagrange equation of the variational problem (compare Exercise 3, Section 2–3):

$$\delta \iint \sqrt{1 + p^2 + q^2}\, dx\, dy = \delta \iint F(z, x, y, p, q)\, dx\, dy = 0.$$

Hence according to the rules of the calculus of variations *

$$\frac{\partial F}{\partial z} - \frac{d}{dx}\frac{\partial F}{\delta p} - \frac{d}{dy}\frac{\partial F}{\partial q} = 0,$$

or

$$r(1 + q^2) - 2pqs + t(1 + p^2) = 0,$$

which is Lagrange's equation of the minimal surfaces (Exercise 3, Section 2–8) and equivalent to the condition $M = 0$.

---

* See F. S. Woods, loc. cit., Section 4–4.

Let us now introduce imaginary elements and admit the isotropic curves as parametric lines. Then $E = 0, G = 0, F \neq 0$, and the condition $M = 0$ becomes

$$Ff = 0 \quad \text{or} \quad f = 0. \tag{4-5}$$

In this case, according to Chapter 3, Eq. (2–7):

$$\Gamma_{12}^1 = \Gamma_{12}^2 = 0,$$

which, according to the Gauss equations (2–6), Chapter 3, shows that $\mathbf{x}_{uv}$ vanishes. The surface is thus a translation surface (Chapter 3, Eq. (2–14)) and its equation can therefore be written

$$\mathbf{x}(u, v) = \mathbf{U}(u) + \mathbf{V}(v), \tag{4-6}$$

where $\mathbf{U} = \mathbf{U}(u)$ and $\mathbf{V} = \mathbf{V}(v)$ are functions of u and v respectively. Moreover, since the curves $u = \text{constant}$, $v = \text{constant}$ are isotropic, the curves $\mathbf{U} = \mathbf{U}(u)$ and $\mathbf{V} = \mathbf{V}(v)$ are isotropic curves. This form (4–6) is due to Monge. Hence we can state this theorem, in the formulation of Lie:

Minimal surfaces can be considered as surfaces of translation, of which the generating curves are isotropic.

In Chapter 1, Eq. (12–8) we have given an explicit expression for the coordinates of an isotropic curve in terms of an arbitrary function. Eq. (4–6) therefore allows us to give an explicit expression for the coordinates of any minimal surface in terms of two arbitrary functions: one, $f(u)$ in u only, the other, $g(v)$, in v only. Such surfaces are, as a rule, imaginary. By selecting the $f(u)$ and $g(v)$ in such a way that $\mathbf{U}(u)$ and $\mathbf{V}(v)$ are conjugate imaginary, we obtain all real minimal surfaces. Their equation can be written as follows (compare with Chapter 1, Eq. (12–8)):

$$\begin{aligned}
x &= (u^2 - 1)f'' - 2uf' + 2f + (\bar{u}^2 - 1)\bar{f}'' - 2\bar{u}\bar{f}' + 2\bar{f}, \\
y &= i[(u^2 + 1)f'' - 2uf' + 2f] - i[(\bar{u}^2 + 1)\bar{f}'' - 2\bar{u}\bar{f}' + 2\bar{f}], \\
z &= 2uf'' - 2f' + 2\bar{u}\bar{f}'' - 2\bar{f}',
\end{aligned} \tag{4-7}$$

where $\bar{u}$ is the complex conjugate of u and $\bar{f}$ of f. When $u = \alpha + i\beta$, $\bar{u} = \alpha - i\beta$,

$$f = \varphi + i\psi, \qquad \bar{f} = \varphi - i\psi.$$

Eq. (4–7) expresses $\mathbf{x}$ as $\mathbf{x}(\alpha, \beta)$. The formulas (4–7) are known as the formulas of *Weierstrass*.

From these equations we can compute the fundamental quantities of the minimal surface. We find (writing for the sake of convenience in notation $\bar{u} = v, \bar{f} = g$):

$$\mathbf{x}_u((u^2 - 1)f''', i(u^2 + 1)f''', 2uf''')$$
$$\mathbf{x}_v((v^2 - 1)g''', i(v^2 + 1)g''', 2vg''')$$
$$\mathbf{x}_{uu}(2uf''' + (u^2 - 1)f^{iv}, 2iuf''' + i(u^2 + 1)f^{iv}, 2f''' + 2uf^{iv})$$
$$\mathbf{x}_{uv}(0, 0, 0)$$
$$\mathbf{x}_{vv}(2vg''' + (v^2 - 1)g^{iv}, 2ivg''' + i(v^2 + 1)g^{iv}, 2g''' + 2vg^{iv}).$$

Hence:

$$E = G = 0, \qquad F = 2(uv + 1)^2 f''' g'''$$
$$\mathbf{N}\left(\frac{u + v}{uv + 1}, i\frac{u - v}{uv + 1}, \frac{-uv + 1}{uv + 1}\right) \qquad (4\text{-}8)$$
$$e = 2f''', \qquad f = 0, \qquad g = 2g'''.$$

The equation of the asymptotic lines is

$$f''' \, du^2 + g''' \, dv^2 = 0,$$

and that of the lines of curvature is

$$f''' \, du^2 - g''' \, dv^2 = 0.$$

These equations show that *both the asymptotic lines and the lines of curvature on a minimal surface can be found by means of quadratures.*

By introducing the lines of curvature as parametric lines by the equations

$$du_1 = \sqrt{f'''} \, du + \sqrt{g'''} \, dv, \qquad i \, dv_1 = \sqrt{f'''} \, du - \sqrt{g'''} \, dv,$$

where we take $\sqrt{g'''}$ as the conjugate imaginary of $\sqrt{f'''}$, then with respect to these real parameters the first differential form becomes, since

$$du_1^2 + dv_1^2 = 4\sqrt{f'''} \sqrt{g'''} \, du \, dv:$$
$$ds^2 = \lambda(du_1^2 + dv_1^2), \qquad \lambda = \lambda(u_1, v_1),$$

and by introducing the asymptotic lines as parametric lines by the equations

$$du_2 = \sqrt{f'''} \, du + i\sqrt{g'''} \, dv, \qquad i \, dv_2 = \sqrt{f'''} \, du - i\sqrt{g'''} \, dv,$$

we obtain the first fundamental form as

$$ds^2 = \mu(du_2^2 + dv_2^2), \qquad \mu = \mu(u_2, v_2).$$

Both the lines of curvature and the asymptotic lines on a minimal surface form an isometric system.

From the expression (4-8) for $\mathbf{N}$ we derive by differentiation that

$$\mathbf{N}_u \cdot \mathbf{N}_u = \mathbf{N}_v \cdot \mathbf{N}_v = 0.$$

This means that, if $d\sigma^2$ is the first fundamental form of the spherical image (Section 2–11, Exercise 6; also p. 156):

$$d\sigma^2 = 2\mathbf{N}_u \cdot \mathbf{N}_v \, du \, dv = 2\rho F \, du \, dv = \rho \, ds^2,$$

where ρ is some function of position:

A minimal surface is mapped conformally on its spherical image. The isotropic curves of the minimal surface are mapped into the isotropic lines of the sphere. And since conformal mapping preserves isometric systems, both the lines of curvature and the asymptotic lines of the minimal surface are mapped into isometric systems on the sphere.

When Eq. (4–6) is replaced by

$$\mathbf{x} = e^{i\alpha}\mathbf{U}(u) + e^{-i\alpha}\mathbf{V}(v), \qquad (4\text{–}9)$$

where α is a constant, then the surfaces (4–9) not only continue to represent minimal surfaces, but also represent real minimal surfaces when $\mathbf{U}$ and $\mathbf{V}$ are conjugate complex. They are called, according to H. A. Schwartz, *associate minimal surfaces.* From Eq. (4–7) we can thus obtain a set of ∞^1 real minimal surfaces by replacing f by $e^{i\alpha}f$ and $g = \bar{f}$ by $e^{-i\alpha}g$. But under this transformation, according to Eq. (4–8), F and $\mathbf{N}$ remain unchanged, as well as $eg - f^2$ and $EG - F^2$. *The ∞^1 surfaces (4–9) are therefore isometric, and at corresponding points the tangent planes are parallel.* They pass into each other by bending.

When we call the associate surface obtained for $\alpha = \pi/2$ the *adjoint* minimal surface,

$$\mathbf{y} = i\mathbf{U} - i\mathbf{V}, \qquad (4\text{–}10)$$

then all associate surfaces $\mathbf{z} = \mathbf{z}(u, v)$ can be expressed as follows in terms of $\mathbf{y}$ and the original surface $\mathbf{x}(\alpha = 0)$:

$$\mathbf{z} = \mathbf{x} \cos \alpha + \mathbf{y} \sin \alpha. \qquad (4\text{–}11)$$

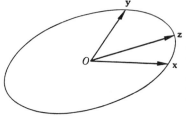

Fig. 5–10

We conclude from this that the end point of the vector $\mathbf{z}$ *describes with varying α an ellipse* of which the end points of the vectors $\mathbf{x}$ and $\mathbf{y}$ mark two vertices (Fig. 5–10). Hence, summing up:

A minimal surface admits a continuous isometric deformation (applicability), in which each point describes an ellipse.

Comparison of the equations of the lines of curvature and of the asymptotic lines shows that in two adjoint minimal surfaces the lines of curvature correspond to the asymptotic lines.

EXAMPLE. Take $f = a(u \ln u - u)$; then

$$f' = a \ln u, \qquad f'' = au^{-1}.$$

If

$$u = -\alpha e^{i\varphi}, \qquad r = 2a(\alpha + \alpha^{-1}),$$

then

$$x = a(\alpha + \alpha^{-1})(e^{i\varphi} + e^{-i\varphi}) = 2a(\alpha + \alpha^{-1}) \cos \varphi = r \cos \varphi,$$

$$y = ai(\alpha + \alpha^{-1})(e^{i\varphi} - e^{-i\varphi}) = 2a(\alpha + \alpha^{-1}) \sin \varphi = r \sin \varphi,$$

$$z = 4 - 4 \ln \alpha, \quad \text{or} \quad r = \frac{b}{2}\left(e^{\frac{z-4}{b}} + e^{-\frac{z-4}{b}}\right), \qquad b = 4a,$$

which is a catenoid.

For the adjoint surface (4–10) we take $r_1 = 2a(\alpha - \alpha^{-1})$. Then

$$x = ia(\alpha - \alpha^{-1})(e^{i\varphi} - e^{-i\varphi}) = r_1 \sin \varphi = r_1 \sin \varphi_1, \qquad (\varphi_1 = \pi - \varphi)$$

$$y = a(\alpha^{-1} - \alpha)(e^{i\varphi} + e^{-i\varphi}) = -r_1 \cos \varphi = r_1 \cos \varphi_1,$$

$$z = -2ia \ln e^{-2i\varphi} = 4a\varphi = (4\pi - 4\varphi_1)a,$$

which is a right helicoid.

We have thus shown that a right helicoid and a catenoid are not only minimal surfaces, but that one surface can pass into the other by a continuous sequence of isometric transformations (see Section 3–5).

Minimal surfaces belong to the best-studied surfaces in differential geometry. Their theory was initiated by Lagrange as an application of his studies in the calculus of variations (1760–1761, *Oeuvres I*, p. 335). Monge, Meusnier, Legendre, Bonnet, Riemann, and Lie contributed to the theory; it was Meusnier who discovered the two "elementary" minimal surfaces, the catenoid and the right helicoid. Karl Weierstrass (*Monatsber. Berlin Akad.*, 1866) and H. A. Schwartz developed the relationship between the theory of complex analytic functions and the real minimal surfaces (see H. A. Schwartz, *Ges. math. abh. I*). A full discussion of the minimal surfaces, including the history, can be found in Darboux' *Leçons I*, pp. 267 ff.

In the theory of capillarity the importance of the minimal surfaces as surfaces of least potential surface energy was illustrated by the experiments of Plateau, *Statique expérimentale et théorique des liquides* (1873), who dipped a wire in the form of a closed space curve into a soap solution and thus realized minimal surfaces as soap films. The problem of Plateau is the problem of determining the minimal surface through a given curve; it has been studied in great generality by J. Douglas. See *Solution of the problem of Plateau*, Trans. Amer. Mathem. Soc. **33**, 1931; also *American Journal Mathem.* **61**, 1939. See for further details R. Courant-H. Robbins, *What is mathematics?* (1941), p. 385; R. Courant, *Acta mathematica* **72** (1940), pp. 51–98. For soap film experiments with minimal surfaces: R. Courant, *Amer. Math. Monthly* **47**, 1940, pp. 167–174.

EXERCISES

1. Show that, apart from the plane, the right helicoid is the only real ruled minimal surface. Hint: Consider the orthogonal trajectories of the rulings as Bertrand curves. (E. Catalan, *Journal de mathém.* **7** (1842), p. 203.)

2. Show that Eq. (4–11) represents an ellipse.

3. Show that the problem of finding all minimal surfaces is identical with the problem of finding two functions $p = p(x, y)$ and $q = q(x, y)$ such that both $p\, dx + q\, dy$ and $(p\, dy - q\, dx)/\sqrt{1 + p^2 + q^2}$ are exact differentials (Lagrange).

4. Show that all the equations of Weierstrass for real minimal surfaces can be cast into the form

$$x = R[(\tau^2 - 1)f'' - 2\tau f' + 2f] = R \int (1 - \tau^2)F(\tau)\, d\tau,$$

$$y = R[i(\tau^2 + 1)f'' - 2i\tau f' + 2if] = R \int i(\tau^2 + 1)F(\tau)\, d\tau,$$

$$z = R[2\tau f'' - 2f'] = R \int 2\tau F(\tau)\, d\tau,$$

where $F(\tau)$ is an analytic function of the complex variable τ, and R indicates the real part; $f''' = F$.

5. Find the value of $F(\tau)$ in Exercise 4 which leads (a) to the right helicoid, (b) to the catenoid.

6. *Minimal surface of Enneper.* Here $F(\tau) = 3$. Show that this surface is algebraic and that its lines of curvature are plane curves of the third degree (Enneper, *Zeitschrift für Mathem. u. Physik.* **9** (1864), p. 108).

7. *Minimal surface of Henneberg.* Here $F(\tau) = 1 - \tau^{-4}$. Show that this surface is algebraic (Henneberg, *Annali di Matem.* **9** (1878), pp. 54–57). This surface is a so-called *one-sided* or *double surface*, which means that without any breach of continuity we can pass from one side of the surface to the other side, as on the Moebius strip.)

8. *Minimal surface of Scherk.* Here $F(\tau) = 2/(1 - \tau^4)$. Show that the cartesian equation of this surface is

$$(\cos x)e^z = \cos y,$$

and show that this surface is also a translation surface with respect to two families of real curves. (Scherk, Crelle's *Journal f. Mathem.* **13** (1835). This surface was the first minimal surface discovered after Meusnier's discovery of the catenoid and the right helicoid.)

9. Show that the only real surfaces which are mapped conformally upon their spherical image are the spheres and the minimal surfaces.

10. *Parallel surfaces.* The locus of the points **y** which are on the normals to the surface S, $\mathbf{x} = \mathbf{x}(u, v)$, at constant distance λ from **x**,

$$\mathbf{y} = \mathbf{x} + \lambda \mathbf{N},$$

is called a parallel surface to S. Show that (a) **N** is the unit surface normal vector of all parallel surfaces, (b) the parallel surfaces of a minimal surface are surfaces for which $R_1 + R_2 = \text{constant},\quad R_1 = \kappa_1^{-1},\quad R_2 = \kappa_2^{-1}$.

5–5 Ruled surfaces. We have occasionally met ruled surfaces in our discussion, but they were always of a particular type, such as developable surfaces, right conoids (Section 2–2), or the sphere as the locus of imaginary lines (Section 2–8). We shall here present a general theory of ruled surfaces, excluding from the beginning the ruled surfaces with imaginary lines.

We have already defined (Section 2–2) a *ruled surface* as a surface generated by the motion of a straight line, its *generating line, generator,* or *ruling.* When the surface is not developable, it is sometimes called a *scroll.* There are ∞^1 generators on a ruled surface. Let $\mathbf{i} = \mathbf{i}(u)$ be the unit vector in the direction of the generating line passing through a point $A(u)$ of an arbitrary nonisotropic curve C lying on the surface, of which the equation is $\mathbf{x} = \mathbf{x}(u)$. A generic point $P(\mathbf{y})$ of the ruled surface is given by the equation

$$\mathbf{y} = \mathbf{x}(u) + v\mathbf{i}(u) = \mathbf{y}(u, v); \qquad \mathbf{i} \cdot \mathbf{i} = 1. \tag{5-1}$$

The directed distance AP is given by the parameter v (Fig. 5–11). The generating lines are the parametric curves $u = \text{constant}$. The vectors

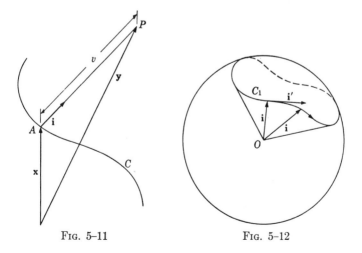

Fig. 5–11   Fig. 5–12

$\mathbf{i} = \mathbf{i}(u)$, drawn through the center O of the unit sphere, describe the *director cone* of the surface (Fig. 5–12); it intersects the sphere in a curve C_1 with equation $\mathbf{i} = \mathbf{i}(u)$, which can be considered as a spherical image of the surface. The curve C on the surface is called its *directrix*.* When $\mathbf{x}$ is a constant the surface is a cone; when $\mathbf{i}$ is a constant the surface is a cylinder.

Our formulas are somewhat simplified if we take as parameter u not the arc length of C, but that of C_1, which is a quantity depending on the nature of the surface only and not on the arbitrary choice of the directrix. This choice of u, and the fact that $\mathbf{i}$ is a unit vector, leads to the identities:

$$\mathbf{i} \cdot \mathbf{i} = \mathbf{i}' \cdot \mathbf{i}' = 1, \qquad \mathbf{i} \cdot \mathbf{i}' = \mathbf{i}' \cdot \mathbf{i}'' = 0.$$

The coefficients of the first and second differential form are obtained from

$$\mathbf{y}_u = \mathbf{x}' + v\mathbf{i}', \qquad \mathbf{y}_v = \mathbf{i}; \qquad \mathbf{i}' = d\mathbf{i}/du, \qquad \mathbf{x}' = d\mathbf{x}/du,$$
$$\mathbf{y}_{uu} = \mathbf{x}'' + v\mathbf{i}'', \qquad \mathbf{y}_{uv} = \mathbf{i}', \qquad \mathbf{y}_{vv} = 0,$$

so that

$$E = \mathbf{x}' \cdot \mathbf{x}' + 2v\mathbf{x}' \cdot \mathbf{i}' + v^2, \qquad F = \mathbf{x}' \cdot \mathbf{i}, \qquad G = 1, \qquad (5\text{–}2)$$
$$De = (\mathbf{x}''\mathbf{x}'\mathbf{i}) + v(\mathbf{i}''\mathbf{x}'\mathbf{i}) + v(\mathbf{x}''\mathbf{i}'\mathbf{i}) + v^2(\mathbf{i}''\mathbf{i}'\mathbf{i}),$$
$$Df = (\mathbf{i}'\mathbf{x}'\mathbf{i}), \qquad g = 0,$$
$$D^2 = EG - F^2 = \mathbf{x}' \cdot \mathbf{x}' - (\mathbf{x}' \cdot \mathbf{i})^2 + 2v\mathbf{x}' \cdot \mathbf{i}' + v^2. \qquad (5\text{–}3)$$

From these equations we derive immediately for the unit normal the expression

$$\mathbf{N} = \frac{\mathbf{x}' \times \mathbf{i} + v\mathbf{i}' \times \mathbf{i}}{D}, \qquad (5\text{–}4)$$

for the Gaussian curvature the expression

$$K = -\frac{f^2}{EG - F^2} = -\frac{(\mathbf{x}'\mathbf{i}\mathbf{i}')^2}{(EG - F^2)^2}, \qquad (5\text{–}5)$$

and for the equation of the asymptotic lines

$$du(e\,du + 2f\,dv) = 0. \qquad (5\text{–}6)$$

The expression (5–4) for the normal vector shows that the tangent plane changes, in general, when its point of tangency moves along a generating

* We speak of the director cone, but of the directrix (curve) because cone is masculine and curve is feminine in Latin. It would really be better in English to use the term *director* in all cases, and then also speak of the Dupin indicator.

line, always passing, of course, through the generating line. We know that **N** is independent of v when the surface is developable. The expression (5–5) for K shows that this is the case when

$$p = (\mathbf{x}'\mathbf{i}\mathbf{i}') = p(u) \tag{5–7}$$

vanishes. This function p is called the *distribution parameter*. We thus have found that *the necessary and sufficient condition that the ruled surface (5–1) be developable is that the distribution parameter vanish.*

From Eq. (5–5) we conclude, since $EG - F^2 > 0$ for real surfaces, that *the Gaussian curvature of real ruled nondevelopable surfaces (scrolls) is negative,* except along those generators where $p(u)$ vanishes.

Eq. (5–6) shows that the straight lines $u = $ constant form one family of asymptotic lines. The other family is given by the equation $e\,du + 2f\,dv = 0$, or

$$\frac{dv}{du} = A + Bv + Cv^2,$$

where A, B, C are certain functions of u. Hence, comparing with Section 1–10, we have found the theorem:

The determination of the curved asymptotic lines of a ruled surface depends on a Riccati equation.

Since the cross ratio of four particular integrals of a Riccati equation is constant, and v is the directed distance, AP, we can immediately conclude that *the cross ratio of the points in which four fixed asymptotic lines intersect the generating lines is constant.*

Further investigation of ruled surfaces is facilitated by the theorem of solid geometry that two generating lines which are not parallel have a common perpendicular.* We exclude here the case that generating lines are parallel, which means that we exclude cylinders. Let us now take two generating lines which are close together, so that they can be given by $u = $ constant, $u + \Delta u = $ constant. Their common perpendicular, as

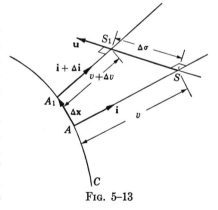

Fig. 5–13

---

* G. Wentworth–D. E. Smith, *Solid Geometry*, Boston; Ginn & Co., 1913, p. 306.

well as the point where it intersects the line u = constant, then assumes a limiting position for $\Delta u \to 0$. To find this position, let us take the generating lines through the points $A(\mathbf{x})$ and $A_1(\mathbf{x} + \Delta \mathbf{x})$ of the directrix C, and let SS_1 be the common perpendicular to the generators through A and A_1. Let vector $\overrightarrow{AS}$ be $v\mathbf{i}$, and vector $\overrightarrow{A_1S_1}$ be $(v + \Delta v)(\mathbf{i} + \Delta \mathbf{i})$, let the unit vector in the direction of SS_1 be $\mathbf{u}$, and the distance SS_1 be $\Delta\sigma$ (Fig. 5–13). The vectors along the sides of quadrilateral AA_1S_1S, if taken with the appropriate sense, add up to zero:

$$\Delta \mathbf{x} + (v + \Delta v)(\mathbf{i} + \Delta \mathbf{i}) - \mathbf{u}\,\Delta\sigma - v\mathbf{i} = 0,$$

or, dividing by Δu and passing to the limit $\Delta u \to 0$:

$$\mathbf{x}' + v\mathbf{i}' + v'\mathbf{i} - \mathbf{u}\sigma' = 0. \tag{5–8}$$

Since $\mathbf{u}$ is perpendicular to $\mathbf{i}$ and to $\mathbf{i} + \Delta \mathbf{i}$, we find that for $\Delta u \to 0$:

$$\mathbf{u} \cdot \mathbf{i} = 0, \qquad \mathbf{u} \cdot \mathbf{i}' = 0,$$

so that ($\mathbf{i}$ and $\mathbf{i}'$ being unit vectors) $\mathbf{u}$ can be taken as

$$\mathbf{u} = \mathbf{i}' \times \mathbf{i}.$$

The vectors $\mathbf{i}, \mathbf{i}', \mathbf{u}$ thus form a set of mutually orthogonal unit vectors. Eq. (5–8) shows how $\mathbf{x}'$ is decomposed in the direction of these vectors; hence

$$\sigma' = \mathbf{u} \cdot \mathbf{x}' = (\mathbf{x}'\mathbf{i}'\mathbf{i}) = -p, \tag{5–9}$$

$$v = -\mathbf{x}' \cdot \mathbf{i}'. \tag{5–10}$$

The third equation, $v' = -\mathbf{x}' \cdot \mathbf{i}$, does not express the derivative of the v of Eq. (5–10) with respect to u, but is the equation of the orthogonal trajectories of the generators $F\,du + G\,dv = 0$ (compare Eq. (5–2)).

Eq. (5–9) gives a new definition of the distribution parameter. When the surface is developable $\sigma' = 0$. This can be expressed by saying that on a developable surface two consecutive generators intersect, or more precisely, that the distance between two generators (u) and $(u + \Delta u)$ is of higher order than Δu.

Eq. (5–10) determines on every generator a certain point S, the *central point*. The tangent plane at that point is called the *central plane;* its unit normal vector is (except when the surface is developable, when the limiting position of $\mathbf{u}$ is perpendicular to the tangent plane):

$$\mathbf{i} \times \mathbf{u} = \mathbf{i} \times (\mathbf{i}' \times \mathbf{i}) = \mathbf{i}'. \tag{5–11}$$

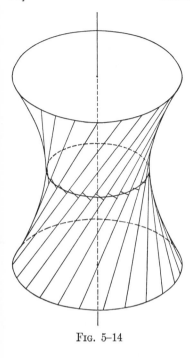

FIG. 5–14

The locus of the points S is called the *striction line* of the surface. On a tangential developable it coincides with the edge of regression, on a cylinder it is indeterminate. Its tangent line does not have the direction of $\mathbf{u}$, as can be seen in Fig. 5–14, which represents a rotation hyperboloid on which the central circle is the striction line. Taking now the striction line as the directrix curve, which is always possible when this line does not reduce to a point (and this happens only when the surface is a cone), then we have the additional condition

$$\mathbf{x}' \cdot \mathbf{i}' = 0.$$

The v in Eq. (5–8) is now zero, since (5–8) relates the point A on the directrix to the central point. Hence (see Eqs. (5–9) and (5–11)):

$$\mathbf{x}' \times \mathbf{i} = (\mathbf{u} \times \mathbf{i})\sigma' = p\mathbf{i}'.$$

Moreover (see Eq. (5–7)):

$$p^2 = (\mathbf{x}'\mathbf{i}\mathbf{i}')^2 = \begin{vmatrix} \mathbf{x}' \cdot \mathbf{x}' & \mathbf{x}' \cdot \mathbf{i} & 0 \\ \mathbf{x}' \cdot \mathbf{i} & 1 & 0 \\ 0 & 0 & 1 \end{vmatrix} = \mathbf{x}' \cdot \mathbf{x}' - (\mathbf{x}' \cdot \mathbf{i})^2. \qquad (5\text{–}12)$$

We can therefore write the unit normal $\mathbf{N}$, according to Eqs. (5–4), (5–3), and (5–11), in the form

$$\mathbf{N} = \frac{p}{\sqrt{p^2 + v^2}}\,\mathbf{i}' + \frac{v}{\sqrt{p^2 + v^2}}\,\mathbf{u}, \qquad (5\text{–}13)$$

where the square root is positive. If φ is the directed angle between the normal vector at a point (v) and at the central point $(v = 0)$ of a generator, we find

$$\cos \varphi = \frac{p}{\sqrt{p^2 + v^2}}, \qquad \sin \varphi = \frac{v}{\sqrt{p^2 + v^2}};$$

hence

$$\tan \varphi = \frac{v}{p}. \qquad (5\text{–}14)$$

This equation expresses the *theorem of Chasles:*

The tangent of the directed angle between a tangent plane at a point P of a generator of a nondevelopable ruled surface and the central plane is proportional to the distance of P to the central point.

When v runs from $-\infty$ to $+\infty$ and $p > 0$, the directed angle φ runs from $-\pi/2$ to $+\pi/2$; when $p < 0$, it runs from $+\pi/2$ to $-\pi/2$. Therefore, when a point moves along a generator, the tangent plane turns 180°. The point where the tangent plane has turned 90° is the central point (hence the name). When $p > 0$ the tangent plane turns counterclockwise, when $p < 0$ it turns clockwise. This allows us to distinguish between *left-handed* and *right-handed* ruled surfaces, respectively. The *asymptotic tangent plane* is perpendicular to the central plane.

For developable surfaces Eq. (5–14) loses its meaning, but from Eq. (5–13) we see that in this case $\mathbf{N} = \mathbf{u}$, independent of v. Eq. (5–11), as already observed, does not hold in this case. Eq. (5–12) here gives

$$\mathbf{x}' \cdot \mathbf{x}' - (\mathbf{x}' \cdot \mathbf{i})^2 = (\mathbf{x}' \times \mathbf{i}) \cdot (\mathbf{x}' \times \mathbf{i}) = 0,$$

which shows (isotropic directrix and isotropic generators have been excluded) that $\mathbf{x}' \times \mathbf{i} = 0$; the generators are tangent to the edge of regression.

Ruled surfaces were investigated first by Monge (in his *Applications*), who established the partial differential equation satisfied by all ruled surfaces (it is of the third order), and then geometrically by Hachette. The present theory is mainly due to F. Minding, *Journal für Mathem.* **18** (1838), pp. 297–302, and M. Chasles, *Corresp. mathém. et phys. de Quetelet* **11** (1839); also to Bonnet. To Chasles we owe the names *central point* and *line of striction*. The theorem on asymptotic lines can be found in a book by Paul Serret, *Théorie nouvelle géométrique et méchanique des courbes à double courbure* (1860).

EXERCISES

1. Find the distribution parameter and the striction line of a right conoid

$$x = v \cos u,$$
$$y = v \sin u, \qquad z = f(u).$$

2. Verify in the case that in Exercise 1 $f(u) = \sqrt{r^2 - a^2 \cos^2 u}$, r and a constants, that $p > 0$ means a left-handed ruled surface, and $p < 0$ a right-handed one. This surface is the *conocuneus (conical wedge) of Wallis* (Fig. 5–15).

3. The normals to a scroll along a generator form a hyperbolic paraboloid.

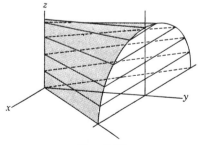

Fig. 5–15

4. The Gaussian curvature of a ruled surface is the same at two points of a generator which are equidistant from the central point.

5. The cross ratio of four points of a generator of a scroll is equal to the cross ratio of the four tangent planes at these points (M. Chasles).

6. Find the asymptotic tangent plane of a generating line of a hyperboloid of one sheet.

7. Show that (a) the striction line of a hyperboloid of revolution is the central circle, (b) the rulings cut it at constant angle, and (c) the parameter of distribution is constant.

8. Show that the striction lines of the hyperbolic paraboloid $\frac{x^2}{a^2} - \frac{y^2}{b^2} = z$ are the parabolas in the planes

$$\frac{x}{a^3} \pm \frac{y}{b^3} = 1.$$

9. Show that the striction line of the hyperboloid

$$\frac{x^2}{a^2} + \frac{y^2}{b^2} - \frac{z^2}{c^2} = 1$$

is the space curve formed by the intersection of the hyperboloid and the surface of the fourth degree,

$$a^2 y^2 z^2 \left(\frac{1}{b^2} + \frac{1}{c^2}\right)^2 + b^2 z^2 x^2 \left(\frac{1}{c^2} + \frac{1}{a^2}\right)^2 - c^2 x^2 y^2 \left(\frac{1}{b^2} - \frac{1}{a^2}\right)^2 = 0.$$

10. Show that the first fundamental form of a ruled surface can be cast into the form $ds^2 = du^2 + ((u - \alpha)^2 - \beta^2) \, dv^2$, where α and β are functions of v alone. Then $u = \alpha$ is the equation of the striction line and

$$K = \frac{-\beta^2}{[(u - \alpha)^2 - \beta^2]^2}.$$

11. Find the equation of the asymptotic tangent plane along a ruling of a ruled surface.

12. The points on a generator of a ruled surface can be paired in such a way that the tangent planes at these points are perpendicular. If P, P_1 form such a pair, and S is the central point, prove that the product $SP \times SP_1$ is constant for all such pairs of points on the same generator.

13. Show that the distribution parameter of a right helicoid is constant. Is the converse true?

14. *A theorem of Bonnet.* If a curve on a ruled surface satisfies any of the three conditions, (a) of being a geodesic, (b) of being a striction line, (c) of intersecting the generators at constant angles, then any two of these conditions implies the third.

15. A space curve is the line of striction on the surface of its binormals, the rectifying plane being the central plane and the tangent the common perpendicular of consecutive binormals. The distribution parameter is $-T$.

5–6 Imaginaries in surface theory. We already have had a few occasions to refer to imaginaries in surface theory. In Section 2–7 we found that asymptotic lines in regions of positive Gaussian curvature are imaginary, and in Section 2–8 we integrated the differential equation of the asymptotic lines for the case of a sphere. The asymptotic lines of a sphere were found to be isotropic lines.

We shall now give a more detailed discussion of isotropic elements on an arbitrary surface. They are defined by the equation $ds^2 = 0$, or

$$E \, du^2 + 2F \, du \, dv + G \, dv^2 = 0. \tag{6-1}$$

When $EG - F^2 \neq 0$ this equation defines two directions, which arrange themselves to the net of *isotropic curves* (also called *minimal curves*). Since the distance of any two points on such curves is zero, they are the lines of shortest real distance between their points, and it therefore seems that they can be considered as geodesics. That this is the case can be shown by introducing the net of isotropic curves as the net of parametric lines on the surface. Then $E = G = 0$, and

$$ds^2 = 2F \, du \, dv, \qquad F = F(u, v), \tag{6-2}$$

so that

$$\Gamma_{11}^2 = \Gamma_{12}^1 = \Gamma_{12}^2 = \Gamma_{22}^1 = 0, \qquad \Gamma_{11}^1 = F_u/F, \qquad \Gamma_{22}^2 = F_v/F, \tag{6-3}$$

and this shows that the equation of the geodesic lines (Chapter 4, (2–3a)) is satisfied for $v = $ constant. By interchanging u and v we also prove that the curves $u = $ constant are geodesics. *The isotropic curves can be considered as geodesics.* We have used the isotropic curves in Section 5–2 for the introduction of isothermic lines and in Section 5–4 for the investigation of minimal surfaces.

Let us now introduce *imaginary surfaces* by considering $\mathbf{x} = \mathbf{x}(u, v)$ as an analytic vector function in two complex variables. We can maintain most of the conceptions of real surface theory by defining them by means of their analytic expressions, provided $EG - F^2 \neq 0$. Thus we can define tangent plane, normal, first and second differential forms, asymptotic lines, lines of curvature, and conjugate lines. Geodesics are defined as curves of zero geodesic curvature, although not of shortest length, except in special cases.

An exception must be made for the case that $EG - F^2 = 0$, when all formulas and definitions in which $EG - F^2$ occur either lose meaning or have to be revised. Because of Eq. (2–8a), Ch. 2, this condition holds for all coordinate systems on the surface, if it holds for one of them, and is therefore a property of the surface itself. Such surfaces, for which at all points

$$EG - F^2 = (\mathbf{x}_u \times \mathbf{x}_v) \cdot (\mathbf{x}_u \times \mathbf{x}_v) = 0, \tag{6-4}$$

are called *isotropic surfaces*. Since ds^2 here is a perfect square, there exists only one family of isotropic curves on these surfaces. The tangent plane exists at all points and is tangent to the local isotropic cone. The *isotropic planes* discussed in Section 1–12 form a special case of isotropic surfaces.

It is convenient to introduce on these surfaces as one set of parametric lines $u = $ constant, the curves for which $ds^2 = 0$. Then ds^2 takes the form:

$$ds^2 = E \, du^2, \qquad E = E(u, v). \tag{6-5}$$

From Eq. (6–5) we derive the following relations:

$$\mathbf{x}_u \cdot \mathbf{x}_u = E, \quad \mathbf{x}_u \cdot \mathbf{x}_v = 0, \quad \mathbf{x}_v \cdot \mathbf{x}_v = 0, \quad \mathbf{x}_{vv} \cdot \mathbf{x}_u = -\mathbf{x}_{uv} \cdot \mathbf{x}_v = 0, \quad \mathbf{x}_v \cdot \mathbf{x}_{vv} = 0.$$

so that we find for $\mathbf{x}_{vv}$ the expression

$$\mathbf{x}_{vv} = \alpha \mathbf{x}_u + \beta \mathbf{x}_v = \beta \mathbf{x}_v, \qquad \beta = \beta(u, v). \tag{6-6}$$

We use here the proposition that when $\mathbf{a} \cdot \mathbf{a} = 0$ and $\mathbf{b} \cdot \mathbf{a} = 0$, then every vector $\mathbf{c}$ for which $\mathbf{c} \cdot \mathbf{a} = 0$ lies in the plane of $\mathbf{a}$ and $\mathbf{b}$. See Exercise 3, Section 1–13.

Eq. (6–6) shows that the isotropic curves $u = $ constant are straight isotropic lines. An isotropic surface is therefore a *ruled surface*. Let us take an arbitrary nonisotropic curve $\mathbf{x} = \mathbf{x}(t)$ on the surface. Then, by defining the isotropic line through every point of the curve by the vector field $\mathbf{u}(t)$, we can write the equation of the surface as follows:

$$\mathbf{y} = \mathbf{x}(t) + \mu \mathbf{u}(t) = \mathbf{y}(\mu, t); \qquad \mathbf{u} \cdot \mathbf{u} = 0, \tag{6-7}$$

where μ is a parameter varying along the isotropic lines. Writing

$$ds^2 = E \, dt^2 + 2F \, dt \, d\mu + G \, d\mu^2,$$

we find that since

$$\mathbf{y}_t = \mathbf{x}' + \mu \mathbf{u}', \qquad \mathbf{y}_\mu = \mathbf{u},$$

the coefficient

$$G = \mathbf{y}_\mu \cdot \mathbf{y}_\mu = 0.$$

From $EG - F^2 = 0$ follows that $F = 0$. Now $F = \mathbf{x}' \cdot \mathbf{u}$, so that the necessary and sufficient condition that Eq. (6–7) represent an isotropic surface is that $\mathbf{x}' \cdot \mathbf{u} = 0$. However, *this is also the condition that the surface* (6–7) *is developable.* We can indeed determine μ as a function of t in such a way that the curve $\mathbf{y}(t)$ of Eq. (6–7) is tangent to the generating lines of the isotropic surface. The tangent vector to this curve must in this case have the direction of $\mathbf{u}$:

$$\mathbf{x}' + \mu \mathbf{u}' + \mu' \mathbf{u} = \lambda \mathbf{u}, \qquad \lambda = \lambda(t).$$

The vectors $\mathbf{x}'$, $\mathbf{u}'$, and $\mathbf{u}$ are coplanar, since $\mathbf{x}' \cdot \mathbf{u} = \mathbf{u}' \cdot \mathbf{u} = \mathbf{u} \cdot \mathbf{u} = 0$. We thus find for μ the value

$$\mu = \frac{\mathbf{x}' \cdot \mathbf{x}'}{\mathbf{x}' \cdot \mathbf{u}'} = -\frac{\mathbf{u}' \cdot \mathbf{x}'}{\mathbf{u}' \cdot \mathbf{u}'},$$

except for the case that $\mathbf{u}' = 0$, when we have a cylinder with isotropic generators.

The identity of the two expressions for μ follows from the fact that $(\mathbf{x}' \times \mathbf{u}') \cdot (\mathbf{x}' \times \mathbf{u}') = 0$ because $\mathbf{x}'$, $\mathbf{u}'$, and $\mathbf{u}$ are coplanar and $(\mathbf{u} \times \mathbf{u}') \cdot (\mathbf{u} \times \mathbf{u}')$ $= (\mathbf{u} \cdot \mathbf{u})(\mathbf{u}' \cdot \mathbf{u}') - (\mathbf{u} \cdot \mathbf{u}')^2 = 0$.

Substitution of this value of μ into Eq. (6–7) gives us the edge of regression of the developable surface (6–7). This curve, having isotropic tangents, is an isotropic curve. When it shrinks to a point, the surface is a cone.

We can express these results in the theorem:

The surfaces for which $EG - F^2 = 0$ are isotropic developables, that is, they are isotropic planes, isotropic cylinders, isotropic cones, or tangent surfaces to isotropic curves.

Let us now suppose that the ruled surface (6–7) is not developable, but is still generated by isotropic lines. Then $\mathbf{x}' \cdot \mathbf{u} \neq 0$, so that $G = 0$, $EG - F^2 = -F^2 \neq 0$. Furthermore:

$$\mathbf{y}_{tt} = \mathbf{x}'' + \mu\mathbf{u}'', \qquad \mathbf{y}_{t\mu} = \mathbf{u}', \quad \mathbf{y}_{\mu\mu} = 0,$$

so that $g = 0$, but $f \neq 0$. (Also $E \neq 0$, $e \neq 0$.)

The equation of the lines of curvature takes the form

$$dt^2(Ef - eF) = 0, \tag{6–8}$$

and the equation (6–3), Chapter 2, for the normal curvature in direction $d\mu/dt$ becomes

$$\kappa = \frac{e\,dt + f\,d\mu}{E\,dt + F\,d\mu}. \tag{6–9}$$

The case $Ef - eF = 0$ leads to surfaces for which κ in all directions is the same, hence to the sphere (Section 3–5), referred to one set of its isotropic lines as parametric curves t = constant. When $Ef - eF \neq 0$ we see from Eq. (6–8) that the surface has *only one set of lines of curvature, the isotropic generators.* Moreover, we find that the equation for the principal curvatures, Chapter 3, Eq. (7–1), now becomes

$$\begin{vmatrix} e - \kappa E & f - \kappa F \\ f - \kappa F & 0 \end{vmatrix} = (f - \kappa F)^2 = 0,$$

which is a perfect square. *Both principal curvatures are equal*, and equal to the normal curvature in the direction of $t = $ constant.

The condition that both curvatures be equal in the real domain leads back to the sphere, since the condition that the two roots of Eq. (7–1) of Chapter 2 are equal is

$$4(EG - F^2)(eg - f^2) - (Eg + eG - 2fF)^2 = 0, \qquad (6\text{–}10)$$

which for $F = 0$ is identical with

$$(Eg - Ge)^2 + 4EGf^2 = 0,$$

or

$$Eg - Ge = 0, \qquad f = 0,$$

the case of an umbilic.

When we admit imaginaries the sphere is not the only possibility, so that we have established the existence of *surfaces for which the two principal curvatures are equal, although not all normal curvatures are equal.* This was the property which led Monge to the discovery of these surfaces. However, Monge, concentrating on real figures, recognized only the one real curve which exists on these surfaces (since on every isotropic line there is one real point). He thus came to the startling result that these surfaces were really curves. "Ce résultat est extraordinaire," he concluded. At present, having been accustomed by the work of Poncelet and Chasles to the free acceptance of imaginaries in geometry, we prefer to summarize as follows:

Those ruled surfaces of which the generating lines are isotropic, and which are not developables, have one set of lines of curvature, the isotropic lines. At all points the two principal curvatures coincide with the curvature in the direction of the isotropic lines.

Monge discovered these surfaces in Chap. 19 of his *Applications*. A detailed study can be found in G. Scheffers, *Anwendung II*, pp. 283–286, 293–295.

EXERCISES

1. Show that the principal curvatures are equal when the fundamental forms I and II have a factor in common.

2. Show that when we can introduce curvilinear coordinates such that $E = 0$, $G = 0$, $e = 0$, the curves $v = $ constant are straight isotropic lines.

3. Show that the case of Exercise 1 leads to the nondevelopable ruled surfaces with isotropic lines.

4. Show that when a surface has two families of straight isotropic lines, the surface is a sphere or a plane.

5. Show that there are two nondevelopable ruled surfaces with isotropic lines on which a given curve is an asymptotic line (G. Scheffers).

6. Show that every isotropic cone is a quadratic cone.

7. The differential equations of the lines of curvature and the asymptotic lines can be written without the denominator $\sqrt{EG - F^2}$ common to e, f, and g. This allows us to define such lines on isotropic developables. Show that on these surfaces all curves can be considered lines of curvature and that the isotropic generators are the asymptotic lines. Also show that all points can be considered umbilics. (F. S. Woods, *Annals of mathem,* **5** (1903–1904), pp. 46–50.)

8. The only surfaces whose element of arc is an exact differential are isotropic planes. (C. L. E. Moore, *Journal Math. and Physics* **4,** 1925, p. 169.)

SOME PROBLEMS AND PROPOSITIONS

1. *Courbure inclinée.* Through every point P of a curve C, $\mathbf{x} = \mathbf{x}(s)$, passes a unit vector of a field $\mathbf{u}_1(s)$. Then we call $d\mathbf{u}_1/ds$ *the curvature vector of C with respect to the field $\mathbf{u}_1$*. If $d\mathbf{u}_1/ds = \kappa_r \mathbf{u}_2$ ($\mathbf{u}_2$ unit vector, κ_r = relative curvature of C with respect to the field $\mathbf{u}_1$), derive the "Frenet formulas" for $d\mathbf{u}_2/ds$ and $d\mathbf{u}_3/ds$, where $\mathbf{u}_3 = \mathbf{u}_1 \times \mathbf{u}_2$. (Fig. 6–1, $\mathbf{u} = \mathbf{u}_1$.)

This curvature was introduced by A. L. Aoust, loc. cit., Section 3–4, who called it *courbure inclinée.* W. C. Graustein, *Trans. Am. Math. Soc.* **36**, 1934, pp. 542–585, calls $d\mathbf{u}_1/ds$ the *associate curvature vector* of $\mathbf{u}_1$ with respect to C.

Now let two congruences of curves be given on a surface with unit tangent vector field $\mathbf{t}_1$, $\mathbf{t}_2$ respectively, and let s_1, s_2 be the respective arc lengths. Show that the projections of $d\mathbf{t}_1/ds_2$ and of $d\mathbf{t}_2/ds_1$ on the surface normal are equal.

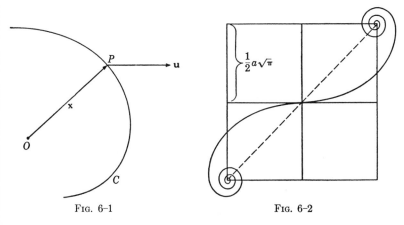

FIG. 6–1   FIG. 6–2

2. *Clothoid.* Find the cartesian equation of the curve with natural equations $Rs = a^2$ (a, a constant) and, taking the point of inflection as origin and the tangent as X-axis, show that the asymptotic points are given by $x = y = (a/2)\sqrt{\pi}$ and $-(a/2)\sqrt{\pi}$. This curve is also called the *spiral of Cornu,* and appears in the theory of diffraction. It was introduced by James Bernoulli (Fig. 6–2). See E. Cesàro, *Natürliche Geometrie,* 1901, p. 15.

3. *Geodesic torsion.* Show that the torsion of the geodesic of a surface with unit tangent vector $\mathbf{t}$ is given by $\mathbf{u} \cdot (d\mathbf{N}/ds)$, where $\mathbf{u}$ is the vector defined in Section 4–1. This quantity is called the *geodesic torsion* τ_g. Show that $\tau_g = 0$ characterizes the lines of curvature.

4. *A moving trihedron on a surface.* If we introduce along a curve $\mathbf{x} = \mathbf{x}(s)$

201

on a surface the trihedron $(\mathbf{t}, \mathbf{u}, \mathbf{N})$ of Problem 3, then Eq. (1–8), Chapter 4, can be complemented as follows:

$$\frac{d\mathbf{t}}{ds} = \kappa_g \mathbf{u} + \kappa_n \mathbf{N}, \qquad \frac{d\mathbf{u}}{ds} = -\kappa_g \mathbf{t} - \tau_g \mathbf{N}, \qquad \frac{d\mathbf{N}}{ds} = -\kappa_n \mathbf{t} + \tau_g \mathbf{u}.$$

Prove from these equations that if two surfaces are tangent to each other along a curve and the normals to them in the points of the curve are similarly directed, the curve has the same geodesic curvature and the same geodesic torsion with respect to both surfaces and the surfaces have the same normal curvature in the direction of the curve. (J. Knoblauch, *Grundlagen der Differentialgeometrie*, Teubner, Leipzig, 1913, p. 56, calls these formulas the *general Frenet formulas* of the theory of curves on surfaces. See also W. C. Graustein, *Differential geometry*, p. 165, and our Exercise 2, Section 4–8, and the Appendix.)

5. *A theorem on Bertrand curves.* If a curve C of constant curvature and a curve C of constant torsion are in such a one-to-one correspondence that the tangent at the corresponding points P, P_1 are parallel, then the locus of the points which divide P, P_1 in a constant ratio is a Bertrand curve. (C. Bioche, *Bull. Soc. math. France* **17**, 1888–89, pp. 109–112. See also A. P. Mellish, loc. cit., Section 1–13.)

6. *Surfaces of constant width.* We define an *ovaloid* as a convex closed surface with continuous nonvanishing principal curvatures $(K > 0)$. Such a surface has two parallel tangent planes in every plane direction (*opposite* tangent planes). When the distance between opposite tangent planes is the same for all directions, we call the ovaloid a *surface of constant width* (see Section 1–13 on ovals and curves of constant width). Prove that for such a surface:

(a) the principal directions at opposite points are equal,

(b) the mean curvatures at opposite points are equal,

(c) its normals are double, that is, the normal at a point is also the normal at the opposite point. (A. P. Mellish, loc. cit., Section 1–13; the theory is due to H. Minkowski, *Ges. Werke II*, pp. 277–279; see also W. Blaschke, *Kreis und Kugel*, Veit, Leipzig, 1916, pp. 138, 150.)

7. Find the expressions (Chapter 1, Eqs. (12–8)) for the isotropic curves by integrating the equation

$$dx^2 + dy^2 = ds^2, \qquad s = iz,$$

by considering x and y as the coordinates of the point of the evolute of a curve C_1. (Introduce as parameter the angle φ of the normal of C_1 with the X-axis, see *Enc. Math. Wiss. III D* 1, 2, p. 26.)

8. *Curvature of asymptotic lines.* When κ_a is the curvature of an asymptotic line l at a point P of a surface, and κ_p that of the branch of the curve of intersection of the surface and the tangential plane at P, then $|\kappa_p| = \frac{2}{3}|\kappa_a|$. (E. Beltrami, **1865**, *Opere matem. I*, p. 255.)

9. *A theorem of Van Kampen.* The tangent t in an asymptotic direction at a point P of a surface of negative curvature is tangent both to the asymptotic curve C and the curve of intersection C_1 of the surface and its tangent plane at S. Then (provided C_1 has its $\kappa \neq 0$ near P) C_1 lies between C and t at P. (E. R. Van Kampen, *Amer. Journal of Mathem.* **61**, 1939, pp. 992–994.)

10. *Natural families of curves.* The extremals of a variational problem

$$\delta \int F \, ds = 0,$$

where F is an arbitrary function of x, y, z, form a *natural family* of curves Through every point passes (in general) one curve of the family in every direction. Show that for these curves the relation holds

$$\mathbf{k} = \mathbf{s} - (\mathbf{s} \cdot \mathbf{t})\mathbf{t}, \qquad \mathbf{s} = \operatorname{grad} \ln F,$$

and derive from this relation that

(a) the centers of the osculating circles of all curves of a natural family which pass through a point P lie in a plane π,

(b) the osculating planes at P pass through the line through P perpendicular to π.

(E. Kasner, *Differential-geometric aspects of dynamics*, Princeton Colloquium, 1909, New York, 1913, 117 pp.).

11. *A theorem of Bonnet on ovaloids.* If the Gaussian curvature of an ovaloid $K \geqslant A^{-2}$, then the maximum distance of two of its points $<\pi A$. For this theorem of O. Bonnet, *Comptes Rendus Acad. Paris* **40**, 1855, pp. 1311–1313, see W. Blaschke, *Differentialgeometrie I*, pp. 218–220.

12. *A theorem of W. Vogt.* If the arc AB of a plane curve has the property that $\kappa(>0)$ decreases from A to B monotonically and if the tangent at A does not meet the arc AB elsewhere, then $\angle TAB > \angle TBA$, if TA and TB are the tangents at A and B and T is on the same side of AB as arc AB. (See Fig. 6–3.) (W. Vogt, *Crelle's Journal für Mathem.* **144**, 1914, pp. 239–248; S. Katsura, *Tôhoku Math. Journ.* **47**, 1940, pp. 94–95.)

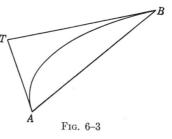

Fig. 6–3

13. *Liouville's Theorem on conformal transformations of space.* The only conformal mappings of space on itself are inversions, similitudes, or a combination of both. (J. Liouville, Note VI to Monge's *Applications*.) One demonstration of this theorem can be given by taking in Section 3–4 $H_1 = H_2 = H_3 = H$. Then $H = k \, (U^2 + V^2 + W^2)^{-1}$, where $U = U(u), V = V(v), W = W(w)$, k constant. (A. R. Forsyth, *Differential geometry*, p. 427.)

14. *W-surfaces.* Surfaces for which there exists a functional relationship $f(\kappa_1, \kappa_2) = 0$ between the principal curvatures are called *Weingarten surfaces* (*W-surfaces*). In this case we can express κ_1 and κ_2 as functions of a parameter w. Show that this can be done in such a way that R_1 and R_2 take the form

$$R_1 = \varphi(w), \qquad R_2 = \varphi(w) - w\varphi'(w),$$

and that the first and third differential forms can be written

$$\mathrm{I} = \left(\frac{\varphi}{w}\right)^2 du^2 + \left(\frac{\varphi - w\varphi'}{\varphi'}\right)^2 dv^2,$$

$$\mathrm{III} = \frac{du^2}{w^2} + \frac{dv^2}{\varphi'^2}.$$

Hint: Use the Codazzi equations (see L. P. Eisenhart, *Differential geometry*, p. 291).

15. *Clothing of surfaces.* Consider a piece of cloth made of threads intersecting at right angles and thus forming a pattern of small squares. Let it be deformed in such a way that it lies smoothly on a given surface ("clothes" the surface). Let us assume that in this process the points of intersection of the threads are not changed, but that the angle at which they intersect may change. Then the threads form on the surface a *net of Čebyšev* and we can introduce a system of curvilinear coordinates such that

$$ds^2 = du^2 + 2 \cos \alpha \, du \, dv + dv^2.$$

Show that

$$K = -\csc \alpha \, \frac{\partial^2 \alpha}{\partial u \, \partial v}.$$

(Tschebycheff, *Sur la coupe des vêtements*, 1878, *Oeuvres II*, p. 708.)

16. *Covariant differential.* Given a vector field $\mathbf{v}(v_i)$, first as function of x, y, z, then as function of orthogonal curvilinear coordinates u, v, w, or u^i, $i = 1, 2, 3$. Show that if $\mathbf{v} = v^1 \mathbf{x}_u + v^2 \mathbf{x}_v + v^3 \mathbf{x}_w$:

$$d\mathbf{v} = \delta v^1 \mathbf{x}_u + \delta v^2 \mathbf{x}_v + \delta v^3 \mathbf{x}_u,$$

where

$$\delta v^i = dv^i + \Gamma^i_{jk} v^j \, du^k \quad (\text{sum on } j, k).$$

The δv^i are called the *covariant differentials* of $\mathbf{v}$ with respect to the system u, v, w. (See for notation Eq. (4–5), Section 3–4.) Also show that the expression for δv^i holds for general curvilinear coordinates in space, the Γ^i_{jk} being defined as in Exercise 14, Section 3–3.

17. *A theorem of Hazzidakis.* On a surface of constant negative curvature $-K$ the asymptotic lines form a net of Čebyšev (Problem 15). For the area A of a quadrangle formed by these asymptotic lines with interior angles α_1, α_2, α_3, α_4 (all $<\pi$) the equation holds

$$KA = 2\pi - \alpha_1 - \alpha_2 - \alpha_3 - \alpha_4.$$

(J. Hazzidakis, *Crelle's Journal für Mathem.* **88**, 1880, pp. 68–73.)

18. *Integral torsion.* The integral torsion of a curve is defined as $\displaystyle\int \tau \, ds.$ This quantity is zero for any closed curve on the sphere, and if on a surface this property holds for all closed curves on it, the surface is a sphere (or a plane). (W. Scherrer, *Vierteljahresschrift Naturforscher Ges. Zürich*, **85**, 1940, pp. 40–46; B. Segre, *Atti Accad. Lincei.* **3**. 1947. pp. 420–426.)

19. *Integral curvature of a curve.* This is defined as $\displaystyle\int |\kappa| \, ds.$ This quantity is $\geqq 2\pi$ for a closed space curve, the sign of equality holding only for plane convex curves (ovals). (W. Fenschel, *Mathem. Annalen*, **101**, 1929, pp. 238–252.)

20. *Umbilics on a closed surface.* On every analytical closed surface of genus zero there exist at least two umbilical points. This conjecture of C. Carathéodory was proved by H. Hamburger, *Acta mathematica* **73**, 1941, pp. 175–332; see G. Bol, *Math. Zeitschr.*, **49**, 1944, pp. 389–410.

APPENDIX

THE METHOD OF PFAFFIANS IN THE THEORY OF CURVES AND SURFACES

1. Pfaffians. A linear differential form of the first order in the differentials $dx_1, dx_2, \ldots, dx_n$ of the variables $x_1, x_2, \ldots, x_n$:

$$v_1 \, dx_1 + v_2 \, dx_2 + \cdots + v_n \, dx_n = \sum v_i \, dx_i, i = 1, 2, \ldots, n,$$

where the v_i are functions of the x_i (which may be constants), is called a *Pfaffian form* or *Pfaffian*. We shall write

$$\omega(d) = v_i \, dx_i \tag{1-1}$$

where (see Section 3–3, Exercise 13) we omit the $\sum$ and agree to sum on the index which is repeated, here i. If no ambiguity exists, we simply write ω instead of $\omega(d)$. The v_i shall be continuous functions with a sufficient number of continuous partial derivatives in the domain X_n (n-dimensional domain with coordinates x_i) considered.

When $n = 3$ the expression $\mathbf{v} \cdot d\mathbf{x}$ is a Pfaffian. In analogy we can call the v_i in Eq. (1–1) the components of the *vector* $\mathbf{v}$ in X_n corresponding to ω.

The following operations with Pfaffians will be useful.

(a) *Linear combination.* If $\omega_1 = v_i \, dx_i$, $\omega_2 = w_i \, dx_i$, $\omega_3 = u_i \, dx_i$, then

$$\lambda\omega_1 + \mu\omega_2 + \nu\omega_3 = (\lambda v_i + \mu w_i + \nu u_i) \, dx_i$$

is a linear combination of the three given Pfaffians. Here the λ, μ, ν are scalar functions which, as in the case of Eq. (5–5), Section 2–5, can be vectors, e.g., $\omega_1 \mathbf{a} + \omega_2 \mathbf{b} = (v_i \mathbf{a} + w_i \mathbf{b}) \, dx_i$. We can construct linear combinations of any number k of Pfaffians.

(b) *Linear dependence.* Such a set of k Pfaffians is *linearly independent* in X_n, if the corresponding vectors are linearly independent. For instance, in the case that $k = 3$, this means that the matrix

$$\begin{bmatrix} v_1 & v_2 & v_3 \\ w_1 & w_2 & w_3 \\ u_1 & u_2 & u_3 \end{bmatrix}$$

is of rank 3. In the case of k vectors the corresponding matrix must be of rank k. In particular, if $n = 2$ and

$$\omega_1 = p_1 \, du + q_1 \, dv, \omega_2 = p_2 \, du + q_2 \, dv, \qquad (1\text{-}2)$$

these two Pfaffians are linearly independent in X_2 if $p_1 q_2 - p_2 q_1 \neq 0$.

  (c) *Exterior multiplication.* When two Pfaffians

$$\omega_1(d) = v_i \, dx_i, \omega_2(\delta) = w_i \, \delta x_i,$$

where dx_i and δx_i represent two different directions in X_n, are combined in the following way:

$$\begin{aligned}
[\omega_1, \omega_2] &= \omega_1(d)\omega_2(\delta) - \omega_2(\delta)\omega_1(d) = (v_i w_j - v_j w_i) \, dx_i \, \delta x_j \\
&= \tfrac{1}{2}(v_i w_j - v_j w_i)(dx_i \, \delta x_j - dx_j \, \delta x_i) \qquad \text{(summed on all } i \text{ and } j) \\
&= (v_i w_j - v_j w_i)(dx_i \, \delta x_j - dx_j \, \delta x_i), \qquad i < j, \qquad (1\text{-}3)
\end{aligned}$$

then we call $[\omega_1, \omega_2]$ the *exterior product* of ω_1 and ω_2. It is alternating, that is

$$[\omega_1, \omega_2] = -[\omega_2, \omega_1], \qquad ([\omega, \omega] = 0). \qquad (1\text{-}4)$$

Necessary and sufficient condition that ω_1 and ω_2 are linearly independent is $[\omega_1, \omega_2] \neq 0$.

  When $n = 2$ and ω_1, ω_2 are given by (1-2):

$$[\omega_1, \omega_2] = (p_1 q_2 - p_2 q_1)(du \, \delta v - dv \, \delta u). \qquad (1\text{-}5)$$

In the special case $\omega_1 = du$, $\omega_2 = \delta v$, we find $[\omega_1, \omega_2] = du \, \delta v - dv \, \delta u$ and we often write $[du \, dv]$ instead of $[du, \delta v]$. Hence

$$[du \, dv] = du \, \delta v - dv \, \delta u,$$

$$[\omega_1, \omega_2] = (v_i w_j - v_j w_i)[dx_i \, dx_j], \qquad i < j.$$

  The vector product of $\mathbf{v} = v_1 \mathbf{e}_1 + v_2 \mathbf{e}_2$ and $\mathbf{w} = w_1 \mathbf{e}_1 + w_2 \mathbf{e}_2$ in the plane can be written in this notation as follows:

$$\mathbf{v} \times \mathbf{w} = (v_1 w_2 - v_2 w_1)\mathbf{e}_1 \times \mathbf{e}_2 = [\mathbf{v}\mathbf{w}] = (v_1 w_2 - v_2 w_1)[\mathbf{e}_1 \mathbf{e}_2].$$

More general, if $\mathbf{v} = v_i \mathbf{e}_i$, $\mathbf{w} = w_i \mathbf{e}_i$ we can define

$$[\mathbf{v}\mathbf{w}] = (v_i w_j - v_j w_i)[\mathbf{e}_i \mathbf{e}_j], \qquad i < j.$$

This is often called the *Grassmann* method.

  (d) *Exterior differentiation.* When $\omega = v_i \, dx_i$, then we define the *exterior derivative* $D\omega$ as follows ($v_{ij} = \partial v_i / \partial x_j$):

$$D\omega = [dv_i, dx_i] = dv_i\,\delta x_j - \delta v_i\,dx_i = (v_{ij} - v_{ji})\,dx_j\,\delta x_i$$

$$\text{(summed on all } i, j\text{)}$$

$$= (v_{ji} - v_{ij})(dx_i\,\delta x_j - dx_j\,\delta x_i) = (v_{ji} - v_{ij})[dx_i\,dx_j], \qquad i < j. \tag{1-6}$$

In particular, if $n = 2, \omega = p\,du + q\,dv$, we find, by virtue of (1-5):

$$D\omega = \left(\frac{\partial q}{\partial u} - \frac{\partial p}{\partial v}\right)[du\,dv]. \tag{1-7}$$

(e) *Some theorems on composition.* It can be readily verified that

$$[\omega_1, \lambda\omega_2 + \mu\omega_3] = \lambda[\omega_1, \omega_2] + \mu[\omega_1, \omega_3], \tag{1-8}$$

$$D(\omega_1 \pm \omega_2) = D\omega_1 \pm D\omega_2, \tag{1-9}$$

$$D(p\omega) = pD\omega + [dp, \omega]. \tag{1-10}$$

$$\text{When } \omega = dp, \quad \text{then} \quad D\omega = 0. \tag{1-11}$$

Here λ, μ, p are scalar functions.

2. Invariance. When in X_n we pass from one system of coordinates x_i to another system x'_i by means of the transformation

$$x_i = f_i(x'_1, x'_2, \ldots, x'_n), \qquad i = 1, 2, \ldots, n,$$

with Jacobian $\neq 0$, so that the x'_i can also be expressed as functions of the x_i, then

$$dx_i = \frac{\partial x_i}{\partial x'_j}\,dx'_j, \qquad dx'_i = \frac{\partial x'_i}{\partial x_j}\,dx_j,$$

and $\omega = v_i\,dx_i$ becomes $\omega = v'_i\,dx'_i$, where

$$v'_i = \frac{\partial x_j}{\partial x'_i}\,v_j.$$

The linear differential form $v_i\,dx_i$ is thus transformed into another linear differential form $v'_i dx'_i$. For the exterior differential forms $\omega_1 = v_i\,dx_i = v'_i\,dx'_i$ and $\omega_2 = w_i\,\delta x_i = w'_i\,\delta x'_i$ we find the exterior product in the new variables in the form

$$[\omega_1, \omega_2] = (v'_i w'_j - v'_j w'_i)[dx'_i\,dx'_j], \qquad i < j,$$

so that this expression also retains its form in the new variables. It is therefore often called the *bilinear covariant* of ω_1 and ω_2. The exterior derivative $D\omega$ shows the same character: $D\omega = (v'_{ji} - v'_{ij})[dx'_i \, dx'_j]$. The notations ω, $[\omega_1, \omega_2]$, $D\omega$ are therefore independent of the coordinates, are *invariant* notations like the vector notation; $\omega = v_i \, dx_i = v'_i \, dx'_i = v''_i \, dx''_i$ etc.

For $n = 2$ (comp. Section 2–1, Eq. 1–11) we find

$$[du \, dv] = \begin{pmatrix} u & v \\ u' & v' \end{pmatrix} [du' \, dv']. \tag{2-1}$$

The expression $\int \omega$, taken along a curve C in X_n, that is, evaluated for the case that the x_i depend on one parameter $x_i = x_i(t)$ is a *line integral*. The notation is an invariant one. We obtain an invariant notation for a *surface integral* $\iint f(u, v) \, du \, dv$ if we change the symbol $du \, dv$ into $[du, dv]$. Indeed, because of (2–1) we find under a change of coordinates

$$\iint f(u, v)[du \, dv] = \iint f'(u', v') \begin{pmatrix} u & v \\ u' & v' \end{pmatrix} [du' \, dv'], \tag{2-2}$$

where $f'(u', v') = f[u'(u, v), v'(u, v)]$ and the integral is taken over a region R in X_2 (a surface). This notation therefore extends to double integrals Leibniz' well-known transformation rule of integrals of functions of one variable:

$$\int f(u) \, du = \int f'(v') \frac{du}{du'} \, du', \qquad f'(v') = f[u(u')].$$

3. Stokes' theorem. By means of Green's theorem (p. 154) an integral along a closed curve C in an X_2 can be transformed into an integral over the surface S enclosed by it. We write the theorem as follows (see Eq. 2–2):

$$\int_C v_1 \, dx_1 + v_2 \, dx_2 = \iint_{S'} \left(\frac{\partial v_2}{\partial x_1} - \frac{\partial v_1}{\partial x_2} \right) [dx_1 \, dx_2]. \tag{3-1}$$

For the orientation see Figure 4–14. We take the region S simply connected and C sufficiently smooth, also in the case of a general X_n, e.g., for $n = 3$, when we have Stokes' theorem as a generalization of Green's theorem. It can be written as follows:

$$\int_C v_1 \, dx_1 + v_2 \, dx_2 + v_3 \, dx_3 = \iint_{S'} \left(\frac{\partial v_2}{\partial x_1} - \frac{\partial v_1}{\partial x_2} \right) [dx_1 \, dx_2]$$

$$+ \left(\frac{\partial v_3}{\partial x_2} - \frac{\partial v_2}{\partial x_3} \right) [dx_2 \, dx_3]$$

$$+ \left(\frac{\partial v_1}{\partial x_3} - \frac{\partial v_3}{\partial x_1} \right) [dx_3 \, dx_1], \tag{3-2}$$

where S is an (orientable) region X_2 bounded by the closed curve C in X_3. If X_3 is ordinary space, $[dx_1\,dx_2]$ can be considered equal to $dA\cos\gamma$ where dA is the element of area of S and γ the angle of its normal with the Z-axis* (compare also Eq. 5–3).

Both Eqs. (3–1) and (3–2) can be written

$$\int_C \omega = \iint_S D\omega. \tag{3–3}$$

It can be shown that this equation holds for any number of variables x_i. It is called *Stokes' theorem* for X_n.

When a function $p(x_i)$ exists such that the v_i are the first partial derivatives of p with respect to the x_i, hence $v_i = \partial p/\partial x_i$, then $\omega = v_i\,dx_i = dp$. We then call ω a *total differential*. In this case $D\omega = 0$ (see Eq. 1–11). Inversely, if $D\omega = 0$ in a region of X_n in which Stokes' theorem holds, then a line integral $\int\omega$ from point A to a point B in X_n is independent of the path and from this can be shown that ω is of the form dp. Under proper safeguards† we can therefore express this property as follows:

Necessary and sufficient condition that ω be a total differential is that $D\omega = 0$.

In ordinary vector analysis this theorem is usually expressed by saying that a vector is a gradient vector if and only if its rotation vanishes.

EXERCISES

1. Find out whether the following Pfaffians (in X_3, resp. X_2) are linearly independent.

(a) $\omega_1 = x_1\,dx_1 + x_2\,dx_2 + x_3\,dx_3$ (b) $\omega_1 = x_1^2\,dx_1 - x_1x_2\,dx_2$
    $\omega_2 = (x_1 + x_2)\,dx_1 - 2x_2\,dx_2$ $\omega_2 = x_1^2x_2\,dx_1 - x_1x_2^2\,dx_2$
    $\omega_3 = dx_1 + x_3\,dx_2 - x_1x_2\,dx_3$

2. Prove that the necessary and sufficient condition that ω_1 and ω_2 are linearly independent is that $[\omega_1, \omega_2] \neq 0$.

3. A set of $n + 1$ Pfaffians in X_n are always linearly dependent. Prove.

4. Let $\omega_1 = x_1\,dx_1 + x_2\,dx_2 + x_3\,dx_3$, $\omega_2 = x_2\,dx_1 - x_1\,dx_2$. Find $[\omega_1, \omega_2]$, $D\omega_1$, and $D\omega_2$.

---

* See, e.g., Ph. Franklin, *A treatise on advanced calculus*, 1940, pp. 380, 382; also R. Creighton Buck, *Advanced calculus*, New York, Toronto, London, 1956, pp. 338, 346.

† See e.g. Buck, *op. cit.*, pp. 356–357.

5. Prove that for directions dx_i and δx_i for which $\delta\,dx_i = d\,\delta x_i$,

$$D\omega = \delta\omega(d) - d\omega(\delta).$$

6. Prove (a): necessary and sufficient condition that for a function $f(x_1, x_2, \ldots, x_n)$ the relation $\delta\,df = d\,\delta f$ holds is $\delta\,dx_i = d\,\delta x_i$;

(b) that in this case $\eta_j\partial\xi_i/\partial x_j = \xi_j\partial\eta_i/\partial x_j$, if the directions dx_i and δx_i are given by $dx_i = \xi_i\,dt$, $\delta x_i = \eta_i\,dt$.

7. Interpreting x and y as cartesian coordinates in the plane, prove by actual integration over a convex region that its area can be expressed as

$$\iint [dx\,dy] = \iint (dx\,\delta y - \delta x\,dy).$$

8. Consider the Pfaffian $\omega = (x\,dy - y\,dx)/(x^2 + y^2)$ in the (not simply connected) region between the concentric circles $x^2 + y^2 = 2$ and $x^2 + y^2 = 4$. Show that $D\omega = 0$ in this region, but that $\int\omega$ is not necessarily zero along a closed path inside this region.

4. Curves in R_3. We consider, as in Section 1–1, a curve C with equation $\mathbf{x} = \mathbf{x}(u)$ in ordinary space with a fixed system of rectangular coordinate axes. To each point of C we associate a trihedron $(\mathbf{e}_1, \mathbf{e}_2, \mathbf{e}_3)$ of mutually orthogonal unit vectors $\mathbf{e}_i$:

$$\mathbf{e}_1\cdot\mathbf{e}_1 = \mathbf{e}_2\cdot\mathbf{e}_2 = \mathbf{e}_3\cdot\mathbf{e}_3 = 1, \quad \mathbf{e}_i\cdot\mathbf{e}_j = 0, \quad i \pm j, \quad i,j = 1,2,3$$

$$(4\text{–}1)$$

(in general not along the coordinates axes). The $\mathbf{e}_i$ are functions of u, and when P moves along C the change of position of P and of the $\mathbf{e}_i$ is given by equations of the form

$$d\mathbf{x} = \omega_i\mathbf{e}_i \qquad d\mathbf{e}_i = \omega_{ij}\mathbf{e}_j, \qquad (4\text{–}2)$$

where the ω_i, ω_{ij} are differentials of the form $f_i(u)\,du$, $f_{ij}(u)\,du$. Since, according to Eq. (4–1) $\mathbf{e}_i\cdot d\mathbf{e}_j = -\mathbf{e}_j\cdot d\mathbf{e}_i = \omega_{ji}$, we find

$$\omega_{ij} = -\omega_{ji} \qquad (\text{hence } \omega_{11} = \omega_{22} = \omega_{33} = 0).$$

We therefore meet in Eq. (4–2) six differentials, $\omega_1, \omega_2, \omega_3, \omega_{12}, \omega_{13}, \omega_{31}$. When we select the trihedron in such a way that $\mathbf{e}_1 = \mathbf{t}, \mathbf{e}_2 = \mathbf{n}, \mathbf{e}_3 = \mathbf{b}$, we obtain in Eq. (4–2) the *Frenet equations* of the curve, with

$$\omega_1 = ds = \sqrt{\dot{\mathbf{x}}_u\cdot\dot{\mathbf{x}}_u}\,du, \ \omega_2 = \omega_3 = 0, \ \omega_{12} = \kappa\,ds, \ \omega_{13} = 0, \ \omega_{23} = \tau\,ds.$$

The *fundamental theorem* (p. 29) states that the ω_i, ω_{ij}, given as single-valued functions of u by means of (4–2), determine one and only one curve but for its position in space, endowed with a trihedron of mutually orthogonal vectors $\mathbf{e}_i$ at each point. For the proof we can follow a reasoning analogous to that given on p. 29.

5. Surfaces in R_3. We now consider a surface S, given by $\mathbf{x} = \mathbf{x}(u, v)$ as in Section 2–1. To each point P of S we associate a (uniquely defined) trihedron $(\mathbf{e}_1, \mathbf{e}_2, \mathbf{e}_3)$ satisfying Eq. (4–1). The $\mathbf{e}_i$ are here functions of u and v, and the change from P in the direction $d\mathbf{x}$ on the surface is given by equations of the form

$$\begin{aligned} d\mathbf{x} &= \omega_i\mathbf{e}_i \\ d\mathbf{e}_i &= \omega_{ij}\mathbf{e}_j \end{aligned}, \qquad \omega_{ij} = -\omega_{ji}, \qquad i, j = 1, 2, 3, \qquad (5\text{--}1)$$

where the six coefficients ω_i, ω_{ij} are now Pfaffians in two variables, e.g. $\omega_1 = p_1(u, v)\, du + q_1(u, v)\, dv$, etc.

We select $\mathbf{e}_3 = \mathbf{N}$ (p. 62). Since $d\mathbf{x} \cdot \mathbf{N} = 0$, the Pfaffian $\omega_3 = 0$, which is the necessary and sufficient condition that $\mathbf{e}_3 = \mathbf{N}$. Since, according to (1–3),

$$\begin{aligned} [\omega_1, \omega_2] &= [\mathbf{e}_1 \cdot d\mathbf{x}, \mathbf{e}_2 \cdot \delta\mathbf{x}] = (\mathbf{e}_1 \times \mathbf{e}_2) \cdot (d\mathbf{x} \times \delta\mathbf{x}) \\ &= \mathbf{e}_3 \cdot (d\mathbf{x} \times \delta\mathbf{x}) = \mathbf{e}_3 \cdot (\mathbf{x}_u \times \mathbf{x}_v)[du\, dv] \\ &= \sqrt{EG - F^2}\, [du\, dv], \end{aligned} \qquad (5\text{--}2)$$

(comp. Eqs. 3–2 and 3–4, Section 2–3), we conclude

$$[\omega_1, \omega_2] = dA \neq 0 \qquad (\sqrt{EG - F^2} > 0, \text{ p. 59}). \qquad (5\text{--}3)$$

In Section 3–3 we have introduced the compatibility relations for Eq. (2–6), Section 3–2, using also the condition $(\mathbf{x}_u)_v = (\mathbf{x}_v)_u$. This leads us to establish the compatibility relations for Eq. (5–1). They express that the components of $d\mathbf{x}$ and $d\mathbf{e}_i$ with respect to the cartesian axes are six total differentials, so that by virtue of the theorem at the end of Section 3, this Appendix:

$$\mathrm{D}\, d\mathbf{x} = 0, \qquad (5\text{--}4)$$

$$\mathrm{D}\, d\mathbf{e}_i = 0. \qquad (5\text{--}5)$$

From Eq. (5–4) follows, using Eqs. (1–8), (1–9) and (1–10),

$$\mathbf{e}_i\mathrm{D}\omega_i + [\mathrm{D}\mathbf{e}_i, \omega_i] = 0, \qquad \text{or}$$

$$\mathbf{e}_i\mathrm{D}\omega_i + [\omega_{ij}, \omega_i]\mathbf{e}_j = \mathbf{e}_i\mathrm{D}\omega_i + [\omega_{ji}, \omega_j]\mathbf{e}_i = 0, \qquad (5\text{--}6)$$

hence

$$\boxed{\mathrm{D}\omega_i = [\omega_{ij}, \omega_j]} \qquad (5\text{--}7)$$

or, since $\omega_3 = 0$:

$$D\omega_1 = [\omega_{12}, \omega_2], \; D\omega_2 = [\omega_{21}, \omega_1] \qquad (5\text{--}7a)$$

$$[\omega_{31}, \omega_1] + [\omega_{32}, \omega_1] = 0.$$

From Eq. (5–5) follows in a similar way

$$\boxed{D\omega_{ij} = [\omega_{jk}, \omega_{ik}] = [\omega_{ik}, \omega_{kj}]} \qquad (5\text{--}8)$$

or

$$D\omega_{23} = [\omega_{21}, \omega_{13}],$$
$$D\omega_{31} = [\omega_{32}, \omega_{21}], \qquad (5\text{--}8a)$$
$$D\omega_{12} = [\omega_{13}, \omega_{32}].$$

Equations (5–7) and (5–8) are the *equations of structure* of the ω_i, ω_{ij}. Eqs. (5–8), as we shall see, are the equivalent of Gauss-Codazzi equations.

The *fundamental theorem* (Section 3–6) states that if the single valued Pfaffians in two variables ω_i, ω_{ij} with $[\omega_1, \omega_2] \neq 0$ satisfy the equations of structure, then they determine one and only one surface, given but for its position in space, and endowed with a trihedron $(\mathbf{e}_1, \mathbf{e}_2, \mathbf{e}_3)$ at each point. For a proof we must refer to the literature, see below.

From Eq. (5–1) we can derive the following expression for the three fundamental forms of the surface (pp. 59, 73, 103):

$$ds^2 = \quad \mathrm{I} = d\mathbf{x} \cdot d\mathbf{x} = (\omega_1)^2 + (\omega_2)^2$$
$$\mathrm{II} = -d\mathbf{x} \cdot \mathbf{e}_3 = \omega_1\omega_{13} + \omega_2\omega_{23}$$
$$\mathrm{III} = d\mathbf{e}_3 \cdot d\mathbf{e}_3 = (\omega_{13})^2 + (\omega_{23})^2.$$

6. Gaussian curvature. We introduce the Gaussian curvature by means of the formula

$$K = \frac{[\omega_{31}, \omega_{32}]}{[\omega_1, \omega_2]}, \qquad (6\text{--}1)$$

in which the denominator is the element of area of the surface (Eq. 5–3): $dA = [\omega_1, \omega_2]$. When dA_S is the element of area of the spherical image $\mathbf{e}_3 = \mathbf{e}_3(u, v)$, Eq. 5–3 shows that $dA_S = [\omega_{13}, \omega_{23}]$, so that Eq. (6–1) is equivalent to $K = dA_S/dA$ (see p. 157). Since

$$[\omega_{13}, \omega_{23}] = [\mathbf{e}_1 \cdot d\mathbf{e}_3, \mathbf{e}_2 \cdot \delta\mathbf{e}_3] = (\mathbf{e}_1 \times \mathbf{e}_2) \cdot (d\mathbf{e}_3 \times \delta\mathbf{e}_3)$$
$$= \mathbf{e}_3 \cdot (\mathbf{N}_u \times \mathbf{N}_v)[du \, dv],$$

we find, with the aid of the Weingarten equations (see p. 156):

$$[\omega_{13}, \omega_{23}] = \frac{eg - f^2}{EG - F^2} \, \mathbf{e}_3 \cdot (\mathbf{x}_u \times \mathbf{x}_v)[du \, dv],$$

from which we derive, using Eq. (5–2):

$$K = \frac{eg - f^2}{EG - F^2}.$$

Now, according to the third of Eq. (5–8a), the numerator of the fraction in Eq. (6–1) can be written as $-D\omega_{12}$; moreover, ω_{12} can be expressed as a linear combination of ω_1 and ω_2 (Exercise 3, Section 3, this Appendix):

$$\omega_{12} = \alpha\omega_1 + \beta\omega_2. \qquad (6-2)$$

By virtue of the first two equations of (5–7a):

$$D\omega_1 = \alpha[\omega_1, \omega_2], \quad D\omega_2 = \beta[\omega_1, \omega_2], \qquad (6-3)$$

so that

$$K = \frac{-1}{[\omega_1, \omega_2]} \, D\left[\frac{D\omega_1}{[\omega_1, \omega_2]} \, \omega_1 + \frac{D\omega_2}{[\omega_1, \omega_2]} \, \omega_2 \right]. \qquad (6-4)$$

The Gaussian curvature can thus be expressed exclusively in terms of ω_1, ω_2, their first and second derivatives. This Eq. (6–4) is equivalent to the *theorema egregium* (p. 111), since bending, a procedure which leaves $ds^2 = (\omega_1)^2 + (\omega_2)^2$ invariant, can be performed in such a way that ω_1 and ω_2 remain themselves invariant (by appropriate selection of the trihedron under this transformation). This theorema egregium has been obtained as a consequence of the third of the equations of structure (5–8a). It can also be shown that the first and second equations of structure (5–8a) are the equivalent of the Codazzi equations (see Exercise 1 of Section 8, this Appendix).

7. Curves on the surface. We have, so far, imposed no special conditions on the set $(\mathbf{e}_1, \mathbf{e}_2)$ except that it be situated in the tangent plane. In this plane, however, we still have ∞^1 choices, depending on one parameter φ:

$$\mathbf{e}_1 = \mathbf{e}_1' \cos \varphi - \mathbf{e}_2' \sin \varphi$$
$$\mathbf{e}_2 = \mathbf{e}_1' \sin \varphi + \mathbf{e}_2' \cos \varphi, \qquad (7-1)$$

which represent rotations about an angle φ, counterclockwise from $(\mathbf{e}_1, \mathbf{e}_2)$

to $(\mathbf{e}_1', \mathbf{e}_2')$ if $\varphi > 0$. The φ is a function of position. Then, if

$$d\mathbf{x} = \omega_i \mathbf{e}_i = \omega_i' \mathbf{e}_i', \quad d\mathbf{e}_i' = \omega_{ij}' \mathbf{e}_j', \quad \mathbf{e}_3 = \mathbf{e}_3' = \mathbf{N}, \qquad (7\text{--}2)$$

we find

$$
\begin{aligned}
d\mathbf{e}_1 &= d(\mathbf{e}_1' \cos \varphi - \mathbf{e}_2' \sin \varphi) \\
&= d\mathbf{e}_1' \cos \varphi - d\mathbf{e}_2' \sin \varphi - \mathbf{e}_1' \sin \varphi \, d\varphi - \mathbf{e}_2' \cos \varphi \, d\varphi \\
&= (\omega_{12}' \mathbf{e}_2' + \omega_{13}' \mathbf{e}_3') \cos \varphi - (\omega_{21}' \mathbf{e}_1' + \omega_{23}' \mathbf{e}_3') \sin \varphi \\
&\quad - \mathbf{e}_1' \sin \varphi \, d\varphi - \mathbf{e}_2' \cos \varphi \, d\varphi,
\end{aligned}
$$

as well as

$$d\mathbf{e}_1 = \omega_{12} \mathbf{e}_2 + \omega_{13} \mathbf{e}_3 = \omega_{12}(\mathbf{e}_1' \sin \varphi + \mathbf{e}_2' \cos \varphi) + \omega_{13} \mathbf{e}_3',$$

so that, equating the coefficients of $\mathbf{e}_1'$, $\mathbf{e}_2'$, $\mathbf{e}_3'$ in both equations, we obtain the transformation formulas

$$\omega_{12}' = \omega_{12} + d\varphi, \qquad \omega_{13} = \omega_{13}' \cos \varphi - \omega_{23}' \sin \varphi.$$

Similar operations on $d\mathbf{x}$ and $\mathbf{e}_2$ give us

$$
\begin{array}{ll}
\text{(a)} \ \omega_1' = \omega_1 \cos \varphi + \omega_2 \sin \varphi & \text{(b)} \ \omega_{31}' = \omega_{31} \cos \varphi + \omega_{32} \sin \varphi \\
\quad \ \omega_2' = -\omega_1 \sin \varphi + \omega_2 \cos \varphi & \quad \ \omega_{32}' = -\omega_{31} \sin \varphi + \omega_{32} \cos \varphi \\
\text{(c)} \ \omega_3' = \omega_3 = 0 & \text{(d)} \ \omega_{12}' = \omega_{12} + d\varphi.
\end{array}
$$

$$(7\text{--}3)$$

These formulas can be used in selecting trihedra in a way appropriate to the study of special curves C on the surface. Let the tangent unit tangent vector $\mathbf{t}$ be given, as on p. 130, by

$$\mathbf{t} = \mathbf{e}_1 \cos \theta + \mathbf{e}_2 \sin \theta, \quad u = u(s), \quad v = v(s), \quad \theta = \theta(s). \quad (7\text{--}4)$$

Then $d\mathbf{x} = \mathbf{t} \, ds = \omega_1 \mathbf{e}_1 + \omega_2 \mathbf{e}_2$, so that along the curve

$$\omega_1 = \cos \theta \, ds, \quad \omega_2 = \sin \theta \, ds, \quad ds = \omega_1 \cos \theta + \omega_2 \sin \theta.$$

We now select $\mathbf{e}_1 = \mathbf{t}$, $\mathbf{e}_2 = \mathbf{u}$ (Section 4–1). With $\mathbf{e}_3 = \mathbf{N}$ this trihedron forms a *natural frame* defined by $\omega_2 = 0$, hence $\omega_1 = ds$. Then, if along the curve $d\mathbf{e}_i = \omega_{ij} \mathbf{e}_j$, $\omega_{ij} = \omega_{ij}(s)$, we can introduce (see Problem 4, p. 202) geodesic curvature, κ_g, normal curvature κ, and geodesic torsion τ_g by the formulas.

$$\omega_{12} = \kappa_g \, ds, \qquad \omega_{13} = \kappa_n \, ds, \qquad \omega_{32} = \tau_g \, ds. \qquad (7\text{--}5)$$

When we replace the natural frame by an arbitrary frame through a rotation θ, in a clockwise direction in order to obtain the relation (7–4), then we obtain for a general frame (with $\mathbf{e}_3 = \mathbf{N}$):

$$\kappa_g \, ds = \omega_{12} + d\theta, \qquad \kappa_n = \omega_{13} \cos \theta + \omega_{23} \sin \theta,$$

$$\tau_g \, ds = \omega_{13} \sin \theta + \omega_{23} \cos \theta. \tag{7–6}$$

The equations of the *geodesics* with respect to an arbitrary frame are obtained in the form $\kappa_g = 0$ or

$$\omega_{12} + d\theta = 0.$$

From Eqs. (7–6), (6–2), and (6–3) we conclude that the geodesics, as well as the geodesic curvature, are bending invariants.

Lines of curvature can be defined as lines of zero geodesic torsion: or

$$\omega_{13} \sin \theta - \omega_{23} \cos \theta = \omega_1 \omega_{23} + \omega_2 \omega_{31} = 0,$$

or

$$\omega_{13} = \kappa \omega_1, \quad \omega_{23} = \kappa \omega_2,$$

which express Rodrigues' theorem (p. 94), with κ as principal curvature.

Asymptotic lines can be defined as lines of zero normal curvature:

$$\omega_{13} \cos \theta + \omega_{23} \sin \theta = \omega_1 \omega_{13} + \omega_2 \omega_{23} = 0.$$

This means that along the asymptotic lines the second fundamental form vanishes.

8. Gauss-Bonnet theorem. Along a closed curve C on the surface, bounding a simply connected region R, as in Fig. 4–15, $\int d\theta = 2\pi$.

Hence, by virtue of Stokes' theorem, Eq. (6–1) for K, and Eq. (7–6) for κ_g:

$$\int \kappa_g \, ds = \int (\omega_{12} + d\theta) = 2\pi + \int \omega_{12} = 2\pi + \int \mathrm{D}\omega_{12}$$

$$= 2\pi + \iint [\omega_{13}, \omega_{23}]$$

$$= -\iint K \, dA + 2\pi,$$

which expresses the Gauss-Bonnet theorem (p. 155).

The method of the moving trihedron and its systematic study by means of Pfaffians is typical of the work of E. Cartan (1869–1951), who, inspired by

S. Lie and G. Darboux, introduced his method c. 1900 for the investigation of continuous groups. See his *Théorie des groupes finis et continus et la géométrie différentielle* (1937). Application to classical differential geometry in W. Blaschke. *Einführung in die Differentialgeometrie* (1950). In Russian: S. P. Finikov, *The method of the exterior forms of Cartan in differential geometry* (1948). For $D\omega = [dv_i \, dx_i]$ we also find the notation $d\omega$ and ω' in the literature.

EXERCISES

1. Prove that the first and second equations of structure (5–8a) are equivalent with the Codazzi equations (3–4), Section 3–3. Take an orthogonal net of curvilinear coordinates or, even more simple, take the lines of curvature as parametric lines.

2. From Eq. (7–6) derive Liouville's formula (1–13, Section 4–1) for the geodesic curvature.

3. From Eq. (6–4) derive Liouville's formula for the Gaussian curvature expressed in Exercise 13, Section 4–2, 2nd formula.

4. Defining developable surfaces as surfaces for which $K = 0$, hence $[\omega_{13}, \omega_{23}] = 0$, show that ds^2 can be written in the form $du^2 + dv^2$.

5. Prove that for two directions on the surface dx_i and dx_i':

$$\omega_1'\omega_{13} + \omega_2'\omega_{23} = \omega_1\omega_{13}' + \omega_2\omega_{23}'$$

and that conjugate lines are characterized by the vanishing of these expressions.

6. Prove that minimal surfaces are characterized by the equations $\omega_{13} = a\omega_2$, $\omega_{23} = a\omega_1$, where a is a scalar function. (Take a natural frame along the asymptotic lines, and use the third equation (5–7a).)

7. Prove that surfaces of constant curvature are characterized by the equation $\omega_1\omega_2 = C\omega_{13}\omega_{23}$, C a constant.

8. A plane curve $\mathbf{x} = \mathbf{x}(s)$ intersects all lines through the origin at an angle of 45°. The moving trihedron is formed by $\mathbf{t}, \mathbf{n}, \mathbf{b}$. Find $\omega_i, \omega_{ij}, i = 1, 2, 3$.

9. On a sphere a trihedron $(\mathbf{e}_1, \mathbf{e}_2, \mathbf{e}_3)$ is defined at each point by the outward surface normal in the direction of $\mathbf{e}_3$ and the vectors $\mathbf{e}_1, \mathbf{e}_2$ along the meridians and parallels in the direction of increasing φ and θ (Fig. 2–1). Find the ω_i, ω_{ij}.

ANSWERS TO PROBLEMS

SECTION 1–6

1. (a) $\kappa^2 = \dfrac{4(9u^4 + 9u^2 + 1)}{(9u^4 + 4u^2 + 1)^3}, \qquad \tau = \dfrac{3}{9u^4 + 9u^2 + 1}.$

   (b) $\kappa^2 = \frac{3}{2}u^6(u^4 + u^2 + 1)^{-3}, \qquad \tau = 0$, curve lies in plane $z = y - x - 1$.

   (c) $\kappa^2 = \dfrac{(f'g'' - f''g')^2 + (f'')^2 + (g'')^2}{[1 + (f')^2 + (g')^2]^3},$

   $\tau = \dfrac{f''g''' - f'''g''}{(f'g'' - f''g')^2 + (f'')^2 + (g'')^2}.$

   (d) $\kappa^2 = a^2(b^2 + 4a^2 \sin^4 u/2)(b^2 + 4a^2 \sin^2 u/2)^{-1},$
   $\tau = -b(b^2 + 4a^2 \sin^4 u/2)^{-1}.$

   (e) $\tau = \dfrac{1}{3a}(1 + u^2)^{-2}.$

2. (a) Differentiate $\mathbf{x} = \lambda \mathbf{t}$.

3. (a) Differentiate $\mathbf{x} = \lambda \mathbf{t} + \mu \mathbf{n}$, or $\mathbf{x} \cdot \mathbf{b} = 0$.

4. $\cos \gamma_3 = \pm a/c$.

5. $d\mathbf{c}/ds = \tau R \mathbf{b} + R'\mathbf{n}$.

6. $\mathbf{c} = -(b^2/a)(\mathbf{e}_1 \cos u + \mathbf{e}_2 \sin u) + bu\mathbf{k}$.

7. If the index 1 refers to C, then $\mathbf{t}_1 = \mathbf{b}$, $ds_1/ds = \tau R$, one more differentiation gives $\kappa_1 \mathbf{n} = -\kappa \mathbf{n}$; select $\mathbf{t} = \mathbf{b}_1$, then differentiation gives $\tau_1 \tau = \kappa^2$.

8. When $(x_0 y_0 z_0)$ is the point, then $x_0 y - y_0 x - b(z - z_0) = 0$ is the plane. The same property holds for all curves for which $x\,dy - y\,dx + k\,dz = 0$, k constant (see E. Goursat, *Cours d'Analyse* I, p. 584).

9. Normal plane: $x \sin 2u + y \cos 2u - z \sin u = 0$. Curve also lies on cylinder $ax + z^2 = a^2$.

10. $\ddot{\mathbf{x}} = t\mathbf{a} + \kappa v^2 \mathbf{n}$, $a = d^2s/dt^2$, $v = ds/dt$.

11. $(\mathbf{x} - \mathbf{c}) \cdot \mathbf{x}''' = 0$.

12. Differentiate $d\varphi/ds = \tan^{-1}(dy/dx)$. Check the sign for a circle.

13. $\mathbf{x} = s\mathbf{t} + \frac{1}{2}s^2(\kappa \mathbf{n}) + \frac{1}{6}s^3(\kappa' \mathbf{n} - \kappa^2 \mathbf{t} + \kappa \tau \mathbf{b}) + \cdots$.

14. $\varphi(u) = c_1 u + c_2$, circular helix.

15. (a) Let $\mathbf{u}$ be the unit vector in direction of common perpendicular at Q and Q_1 to tangents at P and P_1 respectively. Let $PQ = v$, $P_1 Q_1 = v + \Delta v$, $QQ_1 = \Delta \sigma$. Then $\mathbf{u} \cdot \mathbf{t} = 0$, $\mathbf{u} \cdot d\mathbf{t} = 0$, hence $\mathbf{u} = \mathbf{b}$, and $\Delta \mathbf{x} + (\mathbf{t} + \Delta \mathbf{t})(v + \Delta v) - \mathbf{u}\,\Delta \sigma - v\mathbf{t} = 0$, or $v = 0$. Compare the reasoning on p. 191.

   Similarly for (b). Here $\mathbf{u}$ is in direction of $\mathbf{t}$, $v = 0$, hence tangent is common perpendicular. For this method compare Section 5–5.

16. $\mathbf{t}\sqrt{J_1^2 + J_2^2 + J_3^2} = J_1 \mathbf{i} + J_2 \mathbf{j} + J_3 \mathbf{k}$.
    $J_1 = \partial(F_1, F_2)/\partial(y, z)$, $J_2 = \partial(F_1, F_2)/\partial(z, x)$, $J_3 = \partial(F_1, F_2)/\partial(x, y)$.

17. Differentiate $\mathbf{b}_1 = \mathbf{b}$.

19. $ds_t = |d\mathbf{t}| = |d\varphi|$; (a) point, (b) arc of great circle, (c) arc of small circle.

20. $ds_n = |d\mathbf{n}|$, $ds_b = |d\mathbf{b}|$, use Frenet formulas.

Section 1–11

1. $(\mathbf{y} - \mathbf{x}) \times \mathbf{t} = -\frac{1}{2}\kappa s^2 \mathbf{b} + \cdots$.

2. $(\mathbf{y} - \mathbf{x}) \cdot \mathbf{b} = \frac{1}{6}\kappa\tau s^3 + \cdots$.

3. Combine $\mathbf{y} = \mathbf{x} + s\mathbf{x}' + \frac{1}{2}s^2\mathbf{x}'' + \cdots + \dfrac{s^n}{n!}\mathbf{x}^{(n)} + \cdots$ with result of Exercises 1 and 2.

4. Distance to plane is $\pm(\mathbf{a}\cdot\mathbf{x}_1 + p)/\sqrt{\mathbf{a}\cdot\mathbf{a}}$; power with respect to sphere is $(\mathbf{x}_1 - \mathbf{a})\cdot(\mathbf{x}_1 - \mathbf{a}) - p^2$. Power is of order of distance.

5. Helix: $\mathbf{y}_1 = s_1\mathbf{t} + \frac{1}{2}s_1^2\kappa\mathbf{n} + \frac{1}{6}s_1^3(-\kappa^2\mathbf{t} + \kappa\tau\mathbf{b} + \kappa'\mathbf{n}) + \cdots$, κ, τ are constants. Connect points on helix and curve for which $s_1 = s$ (see remark p. 24), then $|\mathbf{y}_1 - \mathbf{y}| = \left|\dfrac{s^3}{6}\kappa'\mathbf{n} + \cdots\right|$. Contact is of order two independent of τ. With method of Exercise 15, Section 1–6, we find $\mathbf{u} = (\tau\mathbf{t} + \kappa\mathbf{b})/\sqrt{\kappa^2 + \tau^2}$, $v^{-1} = (\kappa^2 + \tau^2)/\kappa$.

6. (d) $aR = s^2 + a^2$.

7. $\dfrac{d^3\alpha}{ds^3} + (\kappa^2 + \tau^2)\dfrac{d\alpha}{ds} = 0$, hence $\alpha c = -A\cos cs + B\sin cs + C$,

$c = \sqrt{\kappa^2 + \tau^2}$, select A, B, C appropriately ($A_1 = C_1 = D_1 = 0$, $B_1 = -\kappa$, etc.).

8. Fixed axis direction $\mathbf{a} = \mathbf{e}_1 + \mathbf{e}_3$, cylinder is $y(ab - y)^2 = b^3(x - z)^2$. ($\kappa/\tau = \pm 1$.)

10. See Exercise 5, Section 1–6, $\tau = 0$, $ds_1 = dR$.

11. $\mathbf{x}_1 = \mathbf{x} + a\mathbf{n}$, $\mathbf{n}_1 = \mathbf{n}$.

By differentiation find $a = $ constant. If

$$\mathbf{t}_1 = \mathbf{t}\cos\alpha + \mathbf{b}\sin\alpha, \text{ then } \frac{ds_1}{ds} = \frac{1 - a\kappa}{\cos\alpha} = \frac{a\tau}{\sin\alpha}.$$

By differentiation of $\mathbf{t}_1$ find $\alpha = $ constant.

12. Start with $(1 - a\kappa)\sin\alpha + a\tau\cos\alpha = 0$, α, a constants. Then differentiate $\mathbf{x}_1 = \mathbf{x} + a\mathbf{n}$ to show $\mathbf{n}_1 = \mathbf{n}$.

13. (a) All orthogonal trajectories of the normals (parallel curves) are Bertrand mates; (b) $\mathbf{t}_1 = \mathbf{b}$, $\mathbf{x}_1 = \mathbf{x} + R\mathbf{n}$, compare Exercises 6, 7, Section 1–6; (c) for each a we can find a corresponding α.

Take $\mathbf{u} = \mathbf{t}\sin\alpha - \mathbf{b}\cos\alpha$, $\mathbf{v} = \mathbf{t}\cos\alpha + \mathbf{b}\sin\alpha$, then $\mathbf{v} = \mathbf{u}\times\mathbf{u}'\, a\csc\alpha$ and $\mathbf{t} = \mathbf{u}\sin\alpha + \mathbf{u}\times\mathbf{u}'\, a\cot\alpha$. Result obtained because $a\, d\sigma = (\sin\alpha)\, ds$.

15. From $\mathbf{x}_1 = \mathbf{x} + a\mathbf{n}$, $\mathbf{x} = \mathbf{x}_1 - a\mathbf{n}_1$, $\mathbf{n} = \mathbf{n}_1$, follows $(1 - a\kappa)(1 + a\kappa_1) = \cos^2\alpha$.

16. (a) Curves identical, (b) curves are normal sections of a cylinder.

17. Differentiation gives $\mathbf{t}_1 = \mathbf{t}$, $ds_1/ds = T\kappa$, $\pm\mathbf{n}_1 = \mathbf{n}$, $\mp T\kappa_1 = 1$.

19. From $-dx\sin\theta + dy\cos\theta = ds\sin\alpha$ or $r\, d\theta = ds\sin\alpha$ and $ds^2 = dr^2 + r^2\, d\theta^2 + dz^2$ follows $dz^2 = (r^2\cot^2\alpha\,\theta'^2 - 1)\, dr^2$.

20, 21. Use $x = x(s)$, $y = y(s)$, $z = s\cos\alpha$; the arc length of the base is $ds_1 = ds\sin\alpha$.

22. These helices also intersect the generating lines of the cone at constant angles (loxodromes); their projections on the base intersect all radii at constant angles; from the theorem on the projection of a helix on the plane (Section 1–9) follows that both R and s are proportional to R_1, s_1, and R_1 is proportional to s_1 (Example 2, Section 1–8).

23. If the equation of the paraboloid is $z = ar^2$, then $ar^2 = s\cos\alpha$ and for the projection $ar^2 = s_1\cot\alpha$. This typifies the circle involute (p. 27).

24. $\mathbf{x}'' = \kappa\mathbf{n}$, $\mathbf{x}''' = -\kappa^2\mathbf{t} + \kappa\tau\mathbf{b} - \kappa'\mathbf{n}$, $\mathbf{x}''' \times \mathbf{x}'' = -\kappa^3\left(\mathbf{b} + \dfrac{\tau}{\kappa}\,\mathbf{t}\right)$, this differentiated gives $\mathbf{x}^{iv} \times \mathbf{x}''$, then find $(\mathbf{x}^{iv} \times \mathbf{x}'') \cdot \mathbf{x}'''$.

25. $\mathbf{n} = R\mathbf{x}''$, $-\kappa\mathbf{t} + \tau\mathbf{b} = R\mathbf{x}''' + R'\mathbf{x}''$, by means of this equation express $\mathbf{b}$, and then $\mathbf{b}' = \tau\mathbf{n}$ in terms of $\mathbf{x}'$, $\mathbf{x}''$, $\mathbf{x}'''$, $\mathbf{x}^{iv}$.

Section 1–13

1. Let $y = \pm ix + c_k x^k + c_{k+1}x^{k+1} + \cdots \ (c_k \neq 0)$,
$$a = \sqrt{x^2 + y^2} = \sqrt{\pm 2ic_k x^{(k+1)/2}} + \cdots;$$
$$s = \int_0^x \sqrt{1 + y'^2}\,dx = \frac{2}{k+1}\sqrt{\pm 2ikc_k x^{(k+1)/2}} + \cdots$$
hence $\lim\limits_{x\to 0} s/a = (2\sqrt{k})/k + 1$.

3. Isotropic plane through $\mathbf{a}$ and O is given by $\mathbf{x} \cdot \mathbf{a} = 0$. This property can also be demonstrated geometrically by means of the isotropic circle (at infinity), on which $\mathbf{a}$ is represented by a point P and the plane $\mathbf{x} \cdot \mathbf{a} = 0$ by the tangent through P to the circle.

4. Follows from the fact that $(m - i)/(1 + im) = -i$ is independent of m.

6. Let $A = \dot{\mathbf{x}} \cdot \dot{\mathbf{x}}$, $B = (\dot{\mathbf{x}}\,\ddot{\mathbf{x}}\,\dddot{\mathbf{x}})$, $C = \dot{\mathbf{x}} \times \ddot{\mathbf{x}}$, $D = (\dot{\mathbf{x}} \times \ddot{\mathbf{x}}) \cdot (\dot{\mathbf{x}} \times \ddot{\mathbf{x}})$. Then
(b) $A \neq 0$, $B = 0$, $C \neq 0$, $D \neq 0$;   (c) $A = 0$, $B = 0$, $C = 0$, $D = 0$;
(d) $A = 0$, $B \neq 0$, $C \neq 0$, $D \neq 0$;   (e) $A \neq 0$, $B = 0$, $C \neq 0$, $D = 0$.

7. From $\dot{\mathbf{x}} \cdot \dot{\mathbf{x}} = 0$ follows $(\dot{\mathbf{x}}\,\ddot{\mathbf{x}}\,\dddot{\mathbf{x}})^2 = (\ddot{\mathbf{x}} \cdot \ddot{\mathbf{x}})(\ddot{\mathbf{x}} \cdot \dddot{\mathbf{x}})^3$, hence either $\ddot{\mathbf{x}} \cdot \ddot{\mathbf{x}} = 0$ or $\dot{\mathbf{x}} \cdot \dddot{\mathbf{x}} = 0$, which are equivalent. But $\ddot{\mathbf{x}} \cdot \ddot{\mathbf{x}} = 0$ in (12–8) gives $f''' = 0$, and hence no isotropic (curved) curve. The only possibility is the isotropic straight line.

8. $R_1 + R_2 = (ds/d\varphi) + (ds_1/d\varphi_1) = \mu$.

9. E.g. from $\lambda\lambda' + \mu\mu' = 0$ follows $\lambda = 0$; and when $-d\lambda + \mu\,d\varphi = c\,d\varphi$, then $P = \pi c$, or $c = \mu$ (Exercise 10).

13. Follows from $\kappa_1\,ds = \kappa\,ds$ for evolute and curve, and $\displaystyle\int \mathbf{n}\,ds = \int \mathbf{t}\,ds = \int d\mathbf{x} = 0$ for an oval. The curvature centroid is given by $\mathbf{x}_c = \left(\displaystyle\int \mathbf{x}\kappa\,ds\right)\Big/ \left(\displaystyle\int \kappa\,ds\right) = \dfrac{1}{2\pi}\displaystyle\int \mathbf{x}\kappa\,ds$.

14. Cardioid.

15. The cosine of both angles is $\pm(\mathbf{x} \cdot \mathbf{t})/\sqrt{\mathbf{x} \cdot \mathbf{x}}$.

16. The motion is given by $x' = -x$, $y' = -y + \ln(-1)$.

Section 2–3

2. (a), (b) $u = $ constant, $v = $ constant are straight lines, (c) hyperbolic paraboloid.

3. (a) $ds^2 = (1 + p^2)\,dx^2 + 2\,pq\,dx\,dy + (1 + q^2)\,dy^2$, $\sqrt{1 + p^2 + q^2}\,\mathbf{N} = (-p, -q, 1)$. (b) $ds^2 = dx^2 + dy^2 + dz^2$, $F_x dx + F_y dy + F_z dz = 0$.

5. $r = a\sin\theta$.

6. If $dv/du = \lambda$, use $\lambda_1\lambda_2 = A/C$, $\lambda_1 + \lambda_2 = -2B/C$.

9. $\pm\frac{1}{2}\sqrt{2} = \sqrt{G}\ dv/ds$.

10. Since $r = u\sin\alpha$, the projection is of the form $r = c\exp(k\varphi)$, k constant.

11. $\frac{1}{2}a\pi\sec\alpha$.

12. $x = \frac{1}{2}a(\cos u + \cos v)$, $y = \frac{1}{2}a\,(\sin u + \sin v)$, $z = \frac{1}{2}b(u + v)$, introduce $\frac{1}{2}(u + v)$ as new parameter φ.

13. Eliminate x_1, y_1, z_1 from $\dfrac{xx_1}{a^2} + \dfrac{yy_1}{b^2} + \dfrac{zz_1}{c^2} = 1$, $\dfrac{xa^2}{x_1} = \dfrac{yb^2}{y_1} = \dfrac{zc^2}{z_1} = \dfrac{1}{\lambda}$; $\dfrac{x_1^2}{a^2} + \dfrac{y_1^2}{b^2} + \dfrac{z_1^2}{c^2} = 1$.

SECTION 2–4

1. (a) Tangent surface of $x_1 = u$, $x_2 = u^2$, $x_3 = u^3$. (b) Tangent surface to circular helix.

4. Differentiate $\mathbf{x} = p\mathbf{n} + q\mathbf{b}$ and find that $p^2 + q^2$ is constant.

5. Differentiate $\mathbf{x} = p\mathbf{b} + q\mathbf{t}$ and $\mathbf{x} = p\mathbf{t} + q\mathbf{n}$. The second case is impossible (except for plane curves).

6. Now $\mathbf{a}$ in (4–1) depends on two parameters u, v. Differentiate with respect to u and with respect to v and eliminate u and v. Take $ax + by + cz = 1$, with abc constant.

7. Take $y_3 = 0$ in (4–5), which gives $s^2 = 0$ and $y_2 = \frac{3}{8}\kappa y_1^2$.

8. Take $y_2 = 0$ in (4–5), which gives $s = 0$ and $y_3 = -\dfrac{2\kappa\tau}{3}\,y_1^3$.

10. Take the slope of the line as parameter $-u$. Result: $x = a\cos^3 u$, $y = a\sin^3 u$.

11. Only when the curve is plane (and not a straight line or circle).

12. No.

SECTION 2–8

1. $\mathrm{II} = (r\,dx^2 + 2s\,dx\,dy + t\,dy^2)/\sqrt{1 + p^2 + q^2}$.

4. (a) Right conoid with Z-axis as double line, (b) surfaces of rotation with Z-axis as axis of rotation.

6. $rt - s^2 = 0$.

7. $\kappa_1\cos^2\alpha + \kappa_2\sin^2\alpha + \kappa_1\cos^2\left(\alpha + \dfrac{\pi}{2}\right) + \kappa_2\sin^2\left(\alpha + \dfrac{\pi}{2}\right) = \kappa_1 + \kappa_2$.

8. κ_1 is the curvature of the profile, $|R_2|$ the length of the normal to the profile from profile to axis.

9. $\kappa_1 = 0$, $\kappa_2 = -\tau/v\kappa$.

11. (a) $dr^2 = 2r^2\,d\varphi^2$, etc.

12. All parallel plane sections of an ellipsoid are similar. The circular sections of the surface with equation $\dfrac{x^2}{a^2} + \dfrac{y^2}{b^2} + \dfrac{z^2}{c^2} = 1$ are given by $x^2\left(\dfrac{1}{a^2} - \dfrac{1}{b^2}\right) + z^2\left(\dfrac{1}{c^2} - \dfrac{1}{b^2}\right) = 0$, $(a > b > c)$; the umbilics by $x^2/a^2 = (a^2 - b^2)/(a^2 - c^2)$, $z^2/c^2 = (b^2 - c^2)/(a^2 - c^2)$, $y = 0$.

14. Since **N** is proportional to $F_x\mathbf{e}_1 + F_y\mathbf{e}_2 + F_z\mathbf{e}_3$, the equations are equivalent to $d\mathbf{x} \cdot d\mathbf{N} = 0$, $d\mathbf{x} \cdot \mathbf{N} = 0$.

15. Can be reduced to an equation of the form (6–5b) with $u = x$, $v = y$ by equations of the form $dz = p\,dx + q\,dy$, $F_x + F_z\,p = 0$, $dF_x + F_z(r\,dx + s\,dy) + p\,dF_z = 0$. More elegant deduction by using (**NN′t**) = 0, which follows from Eq. (9–3).

16. This determinant expresses that $(\mathbf{N}\mathbf{N}_x\mathbf{N}_y) = 0$ (see Eq. (8–5)), which can be shown by multiplying the terms of the third row by p and q respectively and adding to those of the first and second row respectively. For a systematic discussion of the theory of curvature of surfaces in the form $F(x, y, z) = 0$, see V and K. Kommerell, *Allgemeine Theorie der Raumkurven und Flächen I*.

17. Eliminate φ from $x = \kappa \tan \varphi \cos \alpha$, $y = \kappa \tan \varphi \sin \alpha$, $z = \kappa$ with the aid of Euler's theorem. The φ is the angle of Eq. (5–11).

SECTION 2–11

1. $\mathbf{y}_{uv} = 0$.

2. $r = cf'$.

3. Follows from $d\mathbf{N}_1 \cdot \mathbf{N}_2 + \mathbf{N}_1 \cdot d\mathbf{N}_2 = 0$ and Rodrigues' theorem.

4. Take $f = F = 0$. Then $\mathrm{I} = E\,du^2 + G\,dv^2$, $\mathrm{II} = E\kappa_1\,du^2 + G\kappa_2\,dv^2$, $\mathrm{III} = E\kappa_1^2\,du^2 + G\kappa_2^2\,dv^2$ (this because of Rodrigues' formula).

5. For asymptotic lines $\mathrm{III} = \tau^2\,ds^2$. For one asymptotic line we have the $+$, and for the other the $-$ sign, but to prove this we must take the asymptotic lines as coordinate lines and compute τ for $u = $ constant and $v = $ constant separately with the aid of the Frenet formulas and Eqs. (2–9) of Chapter 3.

6. (a), (b) Follows from (9–3) and (9–4).

7. Follows from (10–2).

8. (b) Let $r^2 = x^2 + y^2 + z^2$, $du = -\dfrac{r^2}{x^2}\,dx + \dfrac{2r}{x}\,dr$, express $dx^2 + dy^2 + dz^2$ in du^2, dv^2, dw^2.

10. For the roots μ of the equation of Exercise 9 we have $\mu_1 \leqslant a^2 < \mu_2 < b^2$, for those of Eq. (11–5) we have $-\infty < \lambda_1 < \mu_1 < \lambda_2 < \mu_2 < \lambda_3 < \infty$.

SECTION 3–3

5. C is asymptotic because the tangent to C is self-conjugate (C is touched by the curves of two conjugate families of curves, which can be taken as the curves $u = $ constant, $v = $ constant).

6. Parabolas in parallel planes.

7. Write $\dfrac{\partial \ln D}{\partial u} = \dfrac{1}{2D^2}\dfrac{\partial D^2}{\partial u}$.

8. Since $\ln \, \sin \omega = \ln D - \tfrac{1}{2}\ln E - \tfrac{1}{2}\ln G$ (Section 2, Eq. (2–11)) and $\cot \omega = \dfrac{F}{D}$, we find by differentiating $\ln \sin \omega$ the required formulas.

11. Follows from Eqs. (2–9) and (2–6).

12. Take Eq. (2–8a), Chapter 2, and the transformation equation of $eg - f^2$ in Section 2–6, p. 78.

SECTION 3–4

1. $x_u u_x + x_v v_x + x_w w_x = x_x = 1$, etc.

2, 3. These are orthogonality relations of the sets $(\mathbf{u}_1, \mathbf{u}_2, \mathbf{u}_3)$ with respect to $(\mathbf{e}_1, \mathbf{e}_2, \mathbf{e}_3)$ and vice versa (see Section 1–8).

4. $\mathbf{p} = -h_1 \dfrac{dH_1}{ds_2} \mathbf{u}_3 + h_1 \dfrac{dH_1}{ds_3} \mathbf{u}_2$.

SECTION 4–2

1. $ds^2 = dr^2 + r^2\, d\varphi^2,\ \dfrac{d^2\varphi}{ds^2} + \dfrac{2}{r}\dfrac{d\varphi}{ds}\dfrac{dr}{ds} = 0$,

$\dfrac{d\varphi}{ds} = \dfrac{c}{r^2},\ \dfrac{dr}{d\varphi} = \dfrac{r}{c}\sqrt{c^2 - r^2},\ r = a\sec(\varphi + \alpha)$.

2. $ds^2 = a^2(d\theta^2 + \cos^2\theta\, d\varphi^2),\ \dfrac{d^2\varphi}{ds^2} - 2\tan\theta\,\dfrac{d\varphi}{ds}\dfrac{d\theta}{ds} = 0,\ \dfrac{d\varphi}{ds} = c\sec^2\theta$, substitute this into the expression for ds^2.

3. As in 1, with $r = u,\ \varphi \to \varphi\sin\alpha$ (Eq. (1–8), Chapter 2).

4. $(r^2 + a^2)(r^2 + a^2 - c^2)\, d\varphi^2 = c^2\, dr^2$ (Section 2–8).

5. See Exercise 4, Section 3–3.

6. $\mathbf{n} = \pm\mathbf{N}$.

7. According to Example 3, p. 134, $u^2 v' = $ constant, and $\sin\alpha = uv'$.

8. Their osculating planes contain the surface normal ($\mathbf{t}$).

9. From $d\mathbf{x}\cdot d\mathbf{N} = 0$ and $\mathbf{n} = \pm\mathbf{N}$ follows $\kappa = 0$; any straight line on the surface is both asymptotic and geodesic.

10. From $\mathbf{n} = \pm\mathbf{N}$ and $d\mathbf{N} + \kappa\, d\mathbf{x} = 0$ follows $\tau = 0$; (a) and (c) involve (b); (b) and (c) do not involve (a).

11. $D(\mathbf{t}\mathbf{t}'\mathbf{N}) = \left\{ (\mathbf{t}\times\mathbf{t}_u)\dfrac{du}{ds} + (\mathbf{t}\times\mathbf{t}_v)\dfrac{dv}{ds} \right\}\cdot(\mathbf{x}_u\times\mathbf{x}_v);\ \mathbf{t}\times\mathbf{x}_u = -(\mathbf{x}_u\times\mathbf{x}_v)\dfrac{dv}{ds}$,

$\mathbf{t}\times\mathbf{x}_v = +(\mathbf{x}_u\times\mathbf{x}_v)\dfrac{du}{ds};\qquad D(\mathbf{t}\mathbf{t}'\mathbf{N}) = (\mathbf{t}\times\mathbf{t}_u)\cdot(\mathbf{t}\times\mathbf{x}_v) - (\mathbf{t}\times\mathbf{t}_v)\cdot(\mathbf{t}\times\mathbf{x}_u) = \dfrac{\partial}{\partial u}(\mathbf{t}\cdot\mathbf{x}_v) - \dfrac{\partial}{\partial v}(\mathbf{t}\cdot\mathbf{x}_u)$.

12. Take $\mathbf{t}\cdot\mathbf{x}_u = \sqrt{E}\cos\theta,\ \mathbf{t}\cdot\mathbf{x}_v = \sqrt{G}\sin\theta,\ F = 0$.

13. Follows from Chapter 3 (3–7).

14. Follows from Exercise 11.

15. Use Rodrigues' formula.

18. $\displaystyle\int du(U - a)^{-\frac{1}{2}} \pm \int dv(V + a)^{-\frac{1}{2}} = $ constant.

19. For the surfaces of revolution $ds^2 = u^2(du_1^2 + dv^2)$, if $du_1 = \dfrac{du}{u}\sqrt{1 + f'^2}$.

20. Use Eq. (3–8), Chapter 3.

22. This equation expresses that the surface normal lies in the osculating plane.

SECTION 4–8

1. $z = -b(\ln \tan \frac{\varphi}{2} + \cos \varphi), \; 0 < \varphi < \pi.$

2. Follows from (8–3). See also Problem 4, p. 201.

3. The surface normal is the same.

5. Use the Gauss-Bonnet theorem.

6. Take in Fig. 4–17 side A_1A_2 as $v = 0$, A_1A_3 as $v = v_0$, let $u = \varphi(v)$ represent A_2A_3. Then $\iint K \, dA = -\int_0^{v_0} dv \int_0^{\varphi(v)} \frac{\partial^2 \sqrt{G}}{\partial u^2} du = \int_0^{v_0} \left(\frac{\partial \sqrt{G}}{\partial u}\right)_{u=0} dv - \int_0^{v_0} \left(\frac{\partial \sqrt{G}}{\partial u}\right)_{u=\varphi(v)} dv.$ The integrand of the first integral is unity, the second integrand has to be evaluated with the aid of Exercise 4.

11. (c) Check with Exercise 14, Section 4–2.

12. Let $u = $ constant, $v = $ constant be the geodesic distance from $u = 0$, $v = 0$ respectively. Then along the orthogonal trajectories of the parametric lines, for which $F \, du + G \, dv = 0$, $E \, du + F \, dv = 0$, the ds must be du and dv respectively. Hence $E = G = EG - F^2$. Then we obtain from Chapter 2, Eq. (2–11), $E = G = \csc^2 \omega$.

14. Apply Liouville's formula for the case that κ_2 belongs to the orthogonal trajectories of the curves $v = $ constant.

15. The unit normal vector of the rectifying developable is $\mathbf{n}$.

SECTION 5–1

2. (b) $y^2 = 4 \, mx + 4 \, m^2.$

4. $x^2 b^4 - 4y^2 a^4 = 4xa^2 b^2.$

5. If $F \equiv (\mathbf{y} - \mathbf{c}) \cdot (\mathbf{y} - \mathbf{c}) - R^2 = 0$, where $\mathbf{c} = \mathbf{x} + R\mathbf{n}$, then $F_u = 0$ gives $(\mathbf{c} - \mathbf{y}) \cdot \mathbf{n} = R.$

Two osculating circles do not intersect, since the difference $R_1 - R_2$ of their radii is equal to the arc of the evolute between their centers (Exercise 10, Section 1–11) and is therefore longer than the chord connecting them.

7. Differentiate $\mathbf{y} = \mathbf{x} + R\mathbf{n} + \sqrt{a^2 - R^2}\mathbf{b}$ and show that its tangent is tangent to the circle of radius a and center $\mathbf{x}$ in the normal plane.

9. $x^2 + 2y^2 + 2z^2 = 2a^2$ if the given circle is $x^2 + y^2 = a^2.$

10. If S_1 is given by $G = 0$, then (grad F, grad F_u, grad $G) \neq 0$, and $F_x x' + F_y y' + F_z z' = 0$, $F_{ux'} x' + F_{uy'} y' + F_{uz'} z' = 0$, $G_x x' + G_y y' + G_z z' = 0$ at points for which $F_u = 0$, $F_{uu} = 0.$

11. $\mathbf{x}_u \, du + \mathbf{x}_v \, dv + \mathbf{x}_\alpha \, dx = 0$ has only a solution $\neq 0$ if $(\mathbf{x}_u \mathbf{x}_v \mathbf{x}_\alpha) = 0.$

SECTION 5–3

1. Solve the simultaneous equations

$$(E \, du + F \, dv) \, \delta u + (F \, du + G \, dv) \, \delta v = 0,$$
$$(E_1 \, du + F_1 \, dv) \, \delta u + (F_1 \, du + G_1 \, dv) \, \delta v = 0,$$

hence $(E \, du + F \, dv)(F_1 \, du + G_1 \, dv) - (F \, du + G \, dv)(E_1 \, du + F_1 \, dv) = 0$, which has two orthogonal solutions unless E_1, F_1, G_1 are proportional to E, F, G.

2. Introduce geodesic coordinates on both surfaces, then $ds^2 = du^2 + G\,dv^2$ and $ds_1^2 = \sigma^2\,du^2 + \sigma^2 G\,dv^2$, where σ^2 is a function of u alone. Then use Exercise 4, Section 4–8.

4. It means that the second fundamental forms are proportional.

5. Use the Mercator projection.

6. Use Eq. (5–10) and (5–12) of Chapter 4, $\sigma\sqrt{K_1} = \sqrt{K}$.

7. $x = e^{-u}$, $y = v$, use Eq. (2–3a) of Chapter 4, which results in $xy'' - y' - 2(y')^3 = 0$, $y' = dy/dx$.

9. $x = a\varphi$, $y = a\sin\theta$.

11. Take, in accordance with Tissot's theorem, $ds^2 = E\,du^2 + G\,dv^2$, $ds_1^2 = E_1\,du^2 + G_1\,dv^2$; then the equivalence of the geodesic lines (take Eq. (2–1), Section 4–2) gives $\dfrac{\partial}{\partial u}\ln\left(\dfrac{E_1}{E}\dfrac{G^2}{G_1^2}\right) = 0$, $\dfrac{\partial}{\partial v}\ln\left(\dfrac{G_1}{G}\dfrac{E^2}{E_1^2}\right) = 0$, or $E_1 = EU^{-2}V^{-1}$, $G_1 = GU^{-1}V^{-2}$, $U = U(u)$, $V = V(v)$. Also, if $U \neq V$, $\dfrac{\partial \ln E}{\partial v} = \dfrac{\partial \ln (U - V)}{\partial v}$, $\dfrac{\partial \ln G}{\partial u} = \dfrac{\partial \ln (U - V)}{\partial u}$; change of scale on the parametric lines gives the answer.

12. $w = pe^z$.

Section 5–4

4. Follows from Eq. (4–7); the left-hand side of the equations follows from the right-hand side by partial integration.

5. (a) $F(\tau) = ia\tau^{-2}$, (b) $F(\tau) = -a\tau^{-2}$.

6. The equation can be written with appropriate parameter: $x = 3u + 3uv^2 - u^3$, $y = 3v + 3u^2v - v^3$, $z = 3(u^2 - v^2)$; the parametric curves are the lines of curvature. Their planes are given by $x + uz = 3u + 2u^3$, $y - vz = 3v + 2v^3$.

8. It is also a translation surface with real generators $x = u$, $y = 0$, $z = -\ln\cos u$; $x = 0$, $y = v$, $z = \ln\cos v$.

9. Use the formula of Exercise 4, Section 2–11.

10. $\mathbf{N} \cdot \mathbf{y}_u = \mathbf{N} \cdot \mathbf{y}_v = 0$, lines of curvature correspond, Rodrigues' theorem gives $\overline{R}_1 = R_1 - \lambda$, $R_2 = \overline{R}_2 - \lambda$.

Section 5–5

1. Take $\mathbf{y} = f\mathbf{e}_3 + v\mathbf{u}$, $\mathbf{u} = \mathbf{e}_1\cos u + \mathbf{e}_2\sin u$; $p = f'$, Z-axis is striction line.

3. Follows from Eq. (5–14).

4. $K = -p^2/(p^2 + v^2)$, if $\mathbf{x}' \cdot \mathbf{i}' = 0$.

5. Follows from Eq. (5–14) and from $(\tan\varphi_1 - \tan\varphi_2)\cos\varphi_1\cos\varphi_2 = \sin(\varphi_1 - \varphi_2)$.

6. This plane is parallel to the corresponding tangent plane on the asymptotic cone.

7. Write Eq. (5–1) with $\mathbf{x}(a\cos u, a\sin u, 0)$, $\mathbf{i}(\sin u, -\cos u, b)$.

8, 9. Find first the asymptotic tangent plane and then the point of tangency of the perpendicular plane, or find the shortest distance of two generators of the same kind.

11. $(\mathbf{X} - \mathbf{x}, \mathbf{u}\,\mathbf{i}) = 0$.

13. Take one of the helices on the surface as directrix.   Compare Exercise 7.

14. Take $\mathbf{x} = \mathbf{x}(s)$ as the striction line.   Then (a) $\mathbf{x}' \cdot \mathbf{i}' = 0$, (b) $\mathbf{N} = \mathbf{n}$, (c) $\mathbf{i} \cdot \mathbf{t} = $ constant.

15. Take $\mathbf{y} = \mathbf{x} + v\mathbf{b}$, parameter u satisfies $du/ds = \tau$.

Section 5–6

1. The condition is Eq. (6–10). Use e.g. the elimination method of Sylvester.

2. From $E = 0$, $e = 0$ follows $\mathbf{x}_{uu} \cdot \mathbf{N} = 0$, $\mathbf{x}_u \cdot \mathbf{x}_{uu} = 0$, hence $\mathbf{x}_{uu} = \lambda(\mathbf{N} \times \mathbf{x}_v)$; since $G = 0$ this means that $\mathbf{x}_{uu} = \mu\mathbf{x}_u$.

3. Taking $E = 0$, $G = 0$, we can now make $e = 0$, from which we can show (Exercise 2) that the curves $u = $ constant are straight isotropic lines.

4. In this case we can make $E = 0$, $G = 0$, $e = 0$, $g = 0$ and κ is constant in all directions.

5. There are two isotropic directions in the osculating plane of the given curve.

6. Intersect the cone with a plane through the vertex.

7. We can make $F = 0$, $G = 0$, $f = 0$, $g = 0$.

8. Here E in Eq. (6–5) is a function of u alone. Take $E = 1$. Then prove that $z_{uu} = z_{uv} = z_{vv} = 0$, if $x = u$, $y = v$.

APPENDIX

Section 3

1. (a) independent, (b) dependent $(\omega_2 = x_2\omega_1)$

4. $[\omega_1, \omega_2] = -(x_1^2 + x_2^2)[dx_1\,dx_2] + x_2x_3[dx_3\,dx_1] + x_1x_3[dx_2\,dx_3]$,
   $D\omega_1 = 0$, $D\omega_2 = -2[dx_1\,dx_2]$.

5. Follows from: $d\omega(\delta) - \delta\omega(d) = (dv_i)\,\delta x_i + v_i\,d\,\delta x_i - (\delta v_i)\,dx_i - v_i\,\delta\,dx_i = [dv_i\,dx_i] + v_i(\delta\,dx_i - d\,\delta x_i)$.

6. Follows from: $d\,\delta f - \delta\,df = (f_{ji} - f_{ij})[dx_i\,dx_j] + f_i(d\,\delta x_i - \delta\,dx_i)$,
   $f_i = \partial f/\partial x_i$, $f_{ij} = \partial^2 f/\partial x_i\partial x_j$.

8. $\int \omega = 2\pi$ along the circle $x^2 + y^2 = a^2$, $2 < a < 4$.

Section 8

4. From $[\omega_{31}, \omega_{32}] = 0$ follows $D\omega_{12} = 0$, hence ω_{12} can be written $d\varphi$. Take $\varphi = $ const. and the orthogonal trajectories as new coordinates.

8. $\omega_1 = e^\theta\,d\theta\sqrt{2}$, $\omega_2 = 0$, $\omega_{12} = d\theta$, $\omega_{13} = \omega_{23} = 0$.

9. $\omega_1 = d\theta$, $\omega_2 = d\varphi\cos\theta$, $\omega_{12} = -\sin\theta\,d\varphi$, $\omega_{23} = -\cos\theta\,d\varphi$, $\omega_{31} = d\theta$.

INDEX

A CATALOG OF SELECTED
DOVER BOOKS
IN SCIENCE AND MATHEMATICS

A CATALOG OF SELECTED

DOVER BOOKS

IN SCIENCE AND MATHEMATICS

QUALITATIVE THEORY OF DIFFERENTIAL EQUATIONS, V.V. Nemytskii and V.V. Stepanov. Classic graduate-level text by two prominent Soviet mathematicians covers classical differential equations as well as topological dynamics and ergodic theory. Bibliographies. 523pp. 5⅜ x 8½. 65954-2 Pa. $14.95

MATRICES AND LINEAR ALGEBRA, Hans Schneider and George Phillip Barker. Basic textbook covers theory of matrices and its applications to systems of linear equations and related topics such as determinants, eigenvalues and differential equations. Numerous exercises. 432pp. 5⅜ x 8½. 66014-1 Pa. $10.95

QUANTUM THEORY, David Bohm. This advanced undergraduate-level text presents the quantum theory in terms of qualitative and imaginative concepts, followed by specific applications worked out in mathematical detail. Preface. Index. 655pp. 5⅜ x 8½. 65969-0 Pa. $14.95

ATOMIC PHYSICS (8th edition), Max Born. Nobel laureate's lucid treatment of kinetic theory of gases, elementary particles, nuclear atom, wave-corpuscles, atomic structure and spectral lines, much more. Over 40 appendices, bibliography. 495pp. 5⅜ x 8½. 65984-4 Pa. $13.95

ELECTRONIC STRUCTURE AND THE PROPERTIES OF SOLIDS: The Physics of the Chemical Bond, Walter A. Harrison. Innovative text offers basic understanding of the electronic structure of covalent and ionic solids, simple metals, transition metals and their compounds. Problems. 1980 edition. 582pp. 6⅛ x 9¼. 66021-4 Pa. $16.95

BOUNDARY VALUE PROBLEMS OF HEAT CONDUCTION, M. Necati Özisik. Systematic, comprehensive treatment of modern mathematical methods of solving problems in heat conduction and diffusion. Numerous examples and problems. Selected references. Appendices. 505pp. 5⅜ x 8½. 65990-9 Pa. $12.95

A SHORT HISTORY OF CHEMISTRY (3rd edition), J.R. Partington. Classic exposition explores origins of chemistry, alchemy, early medical chemistry, nature of atmosphere, theory of valency, laws and structure of atomic theory, much more. 428pp. 5⅜ x 8½. (Available in U.S. only) 65977-1 Pa. $11.95

A HISTORY OF ASTRONOMY, A. Pannekoek. Well-balanced, carefully reasoned study covers such topics as Ptolemaic theory, work of Copernicus, Kepler, Newton, Eddington's work on stars, much more. Illustrated. References. 521pp. 5⅜ x 8½. 65994-1 Pa. $12.95

PRINCIPLES OF METEOROLOGICAL ANALYSIS, Walter J. Saucier. Highly respected, abundantly illustrated classic reviews atmospheric variables, hydrostatics, static stability, various analyses (scalar, cross-section, isobaric, isentropic, more). For intermediate meteorology students. 454pp. 6⅛ x 9¼. 65979-8 Pa. $14.95

RELATIVITY, THERMODYNAMICS AND COSMOLOGY, Richard C. Tolman. Landmark study extends thermodynamics to special, general relativity; also applications of relativistic mechanics, thermodynamics to cosmological models. 501pp. 5⅜ x 8½. 65383-8 Pa. $13.95

APPLIED ANALYSIS, Cornelius Lanczos. Classic work on analysis and design of finite processes for approximating solution of analytical problems. Algebraic equations, matrices, harmonic analysis, quadrature methods, much more. 559pp. 5⅜ x 8½. 65656-X Pa. $13.95

INTRODUCTION TO ANALYSIS, Maxwell Rosenlicht. Unusually clear, accessible coverage of set theory, real number system, metric spaces, continuous functions, Riemann integration, multiple integrals, more. Wide range of problems. Undergraduate level. Bibliography. 254pp. 5⅜ x 8½. 65038-3 Pa. $8.95

INTRODUCTION TO QUANTUM MECHANICS With Applications to Chemistry, Linus Pauling & E. Bright Wilson, Jr. Classic undergraduate text by Nobel Prize winner applies quantum mechanics to chemical and physical problems. Numerous tables and figures enhance the text. Chapter bibliographies. Appendices. Index. 468pp. 5⅜ x 8½. 64871-0 Pa. $12.95

ASYMPTOTIC EXPANSIONS OF INTEGRALS, Norman Bleistein & Richard A. Handelsman. Best introduction to important field with applications in a variety of scientific disciplines. New preface. Problems. Diagrams. Tables. Bibliography. Index. 448pp. 5⅜ x 8½. 65082-0 Pa. $12.95

MATHEMATICS APPLIED TO CONTINUUM MECHANICS, Lee A. Segel. Analyzes models of fluid flow and solid deformation. For upper-level math, science and engineering students. 608pp. 5⅜ x 8½. 65369-2 Pa. $14.95

ELEMENTS OF REAL ANALYSIS, David A. Sprecher. Classic text covers fundamental concepts, real number system, point sets, functions of a real variable, Fourier series, much more. Over 500 exercises. 352pp. 5⅜ x 8½. 65385-4 Pa. $11.95

PHYSICAL PRINCIPLES OF THE QUANTUM THEORY, Werner Heisenberg. Nobel Laureate discusses quantum theory, uncertainty, wave mechanics, work of Dirac, Schroedinger, Compton, Wilson, Einstein, etc. 184pp. 5⅜ x 8½. 60113-7 Pa. $6.95

INTRODUCTORY REAL ANALYSIS, A.N. Kolmogorov, S.V. Fomin. Translated by Richard A. Silverman. Self-contained, evenly paced introduction to real and functional analysis. Some 350 problems. 403pp. 5⅜ x 8½. 61226-0 Pa. $10.95

PROBLEMS AND SOLUTIONS IN QUANTUM CHEMISTRY AND PHYSICS, Charles S. Johnson, Jr. and Lee G. Pedersen. Unusually varied problems, detailed solutions in coverage of quantum mechanics, wave mechanics, angular momentum, molecular spectroscopy, scattering theory, more. 280 problems plus 139 supplementary exercises. 430pp. 6½ x 9¼. 65236-X Pa. $13.95

ASYMPTOTIC METHODS IN ANALYSIS, N.G. de Bruijn. An inexpensive, comprehensive guide to asymptotic methods–the pioneering work that teaches by explaining worked examples in detail. Index. 224pp. 5⅜ x 8½. 64221-6 Pa. $7.95

OPTICAL RESONANCE AND TWO-LEVEL ATOMS, L. Allen and J. H. Eberly. Clear, comprehensive introduction to basic principles behind all quantum optical resonance phenomena. 53 illustrations. Preface. Index. 256pp. 5⅜ x 8½. 65533-4 Pa. $8.95

COMPLEX VARIABLES, Francis J. Flanigan. Unusual approach, delaying complex algebra till harmonic functions have been analyzed from real variable viewpoint. Includes problems with answers. 364pp. 5⅜ x 8½. 61388-7 Pa. $9.95

ATOMIC SPECTRA AND ATOMIC STRUCTURE, Gerhard Herzberg. One of best introductions; especially for specialist in other fields. Treatment is physical rather than mathematical. 80 illustrations. 257pp. 5⅜ x 8½. 60115-3 Pa. $7.95

APPLIED COMPLEX VARIABLES, John W. Dettman. Step-by-step coverage of fundamentals of analytic function theory–plus lucid exposition of five important applications: Potential Theory; Ordinary Differential Equations; Fourier Transforms; Laplace Transforms; Asymptotic Expansions. 66 figures. Exercises at chapter ends. 512pp. 5⅜ x 8½. 64670-X Pa. $12.95

ULTRASONIC ABSORPTION: An Introduction to the Theory of Sound Absorption and Dispersion in Gases, Liquids and Solids, A.B. Bhatia. Standard reference in the field provides a clear, systematically organized introductory review of fundamental concepts for advanced graduate students, research workers. Numerous diagrams. Bibliography. 440pp. 5⅜ x 8½. 64917-2 Pa. $11.95

UNBOUNDED LINEAR OPERATORS: Theory and Applications, Seymour Goldberg. Classic presents systematic treatment of the theory of unbounded linear operators in normed linear spaces with applications to differential equations. Bibliography. I99pp. 5⅜ x 8½. 64830-3 Pa. $7.95

LIGHT SCATTERING BY SMALL PARTICLES, H.C. van de Hulst. Comprehensive treatment including full range of useful approximation methods for researchers in chemistry, meteorology and astronomy. 44 illustrations. 470pp. 5⅜ x 8½. 64228-3 Pa. $12.95

CONFORMAL MAPPING ON RIEMANN SURFACES, Harvey Cohn. Lucid, insightful book presents ideal coverage of subject. 334 exercises make book perfect for self-study. 55 figures. 352pp. 5⅝ x 8¼. 64025-6 Pa. $11.95

OPTICKS, Sir Isaac Newton. Newton's own experiments with spectroscopy, colors, lenses, reflection, refraction, etc., in language the layman can follow. Foreword by Albert Einstein. 532pp. 5⅜ x 8½. 60205-2 Pa. $12.95

GENERALIZED INTEGRAL TRANSFORMATIONS, A.H. Zemanian. Graduate-level study of recent generalizations of the Laplace, Mellin, Hankel, K. Weierstrass, convolution and other simple transformations. Bibliography. 320pp. 5⅜ x 8½. 65375-7 Pa. $8.95

THE ELECTROMAGNETIC FIELD, Albert Shadowitz. Comprehensive undergraduate text covers basics of electric and magnetic fields, builds up to electromagnetic theory. Also related topics, including relativity. Over 900 problems. 768pp. 5⅜ x 8¼. 65660-8 Pa. $18.95

FOURIER SERIES, Georgi P. Tolstov. Translated by Richard A. Silverman. A valuable addition to the literature on the subject, moving clearly from subject to subject and theorem to theorem. 107 problems, answers. 336pp. 5⅜ x 8½. 63317-9 Pa. $9.95

THEORY OF ELECTROMAGNETIC WAVE PROPAGATION, Charles Herach Papas. Graduate-level study discusses the Maxwell field equations, radiation from wire antennas, the Doppler effect and more. xiii + 244pp. 5⅜ x 8½. 65678-0 Pa. $6.95

DISTRIBUTION THEORY AND TRANSFORM ANALYSIS: An Introduction to Generalized Functions, with Applications, A.H. Zemanian. Provides basics of distribution theory, describes generalized Fourier and Laplace transformations. Numerous problems. 384pp. 5⅜ x 8½. 65479-6 Pa. $11.95

THE PHYSICS OF WAVES, William C. Elmore and Mark A. Heald. Unique overview of classical wave theory. Acoustics, optics, electromagnetic radiation, more. Ideal as classroom text or for self-study. Problems. 477pp. 5⅜ x 8½. 64926-1 Pa. $13.95

CALCULUS OF VARIATIONS WITH APPLICATIONS, George M. Ewing. Applications-oriented introduction to variational theory develops insight and promotes understanding of specialized books, research papers. Suitable for advanced undergraduate/graduate students as primary, supplementary text. 352pp. 5⅜ x 8½. 64856-7 Pa. $9.95

A TREATISE ON ELECTRICITY AND MAGNETISM, James Clerk Maxwell. Important foundation work of modern physics. Brings to final form Maxwell's theory of electromagnetism and rigorously derives his general equations of field theory. 1,084pp. 5⅜ x 8½. 60636-8, 60637-6 Pa., Two-vol. set $25.90

AN INTRODUCTION TO THE CALCULUS OF VARIATIONS, Charles Fox. Graduate-level text covers variations of an integral, isoperimetrical problems, least action, special relativity, approximations, more. References. 279pp. 5⅜ x 8½. 65499-0 Pa. $8.95

HYDRODYNAMIC AND HYDROMAGNETIC STABILITY, S. Chandrasekhar. Lucid examination of the Rayleigh-Benard problem; clear coverage of the theory of instabilities causing convection. 704pp. 5⅜ x 8¼. 64071-X Pa. $14.95

CALCULUS OF VARIATIONS, Robert Weinstock. Basic introduction covering isoperimetric problems, theory of elasticity, quantum mechanics, electrostatics, etc. Exercises throughout. 326pp. 5⅜ x 8½. 63069-2 Pa. $9.95

DYNAMICS OF FLUIDS IN POROUS MEDIA, Jacob Bear. For advanced students of ground water hydrology, soil mechanics and physics, drainage and irrigation engineering and more. 335 illustrations. Exercises, with answers. 784pp. 6⅛ x 9¼. 65675-6 Pa. $19.95

NUMERICAL METHODS FOR SCIENTISTS AND ENGINEERS, Richard Hamming. Classic text stresses frequency approach in coverage of algorithms, polynomial approximation, Fourier approximation, exponential approximation, other topics. Revised and enlarged 2nd edition. 721pp. 5⅜ x 8½.   65241-6 Pa. $15.95

THEORETICAL SOLID STATE PHYSICS, Vol. 1: Perfect Lattices in Equilibrium; Vol. II: Non-Equilibrium and Disorder, William Jones and Norman H. March. Monumental reference work covers fundamental theory of equilibrium properties of perfect crystalline solids, non-equilibrium properties, defects and disordered systems. Appendices. Problems. Preface. Diagrams. Index. Bibliography. Total of 1,301pp. 5⅜ x 8½. Two volumes.   Vol. I: 65015-4 Pa. $16.95
Vol. II: 65016-2 Pa. $16.95

OPTIMIZATION THEORY WITH APPLICATIONS, Donald A. Pierre. Broad spectrum approach to important topic. Classical theory of minima and maxima, calculus of variations, simplex technique and linear programming, more. Many problems, examples. 640pp. 5⅜ x 8½.   65205-X Pa. $16.95

THE CONTINUUM: A Critical Examination of the Foundation of Analysis, Hermann Weyl. Classic of 20th-century foundational research deals with the conceptual problem posed by the continuum. 156pp. 5⅜ x 8½.   67982-9 Pa. $6.95

ESSAYS ON THE THEORY OF NUMBERS, Richard Dedekind. Two classic essays by great German mathematician: on the theory of irrational numbers; and on transfinite numbers and properties of natural numbers. 115pp. 5⅜ x 8½.
21010-3 Pa. $5.95

THE FUNCTIONS OF MATHEMATICAL PHYSICS, Harry Hochstadt. Comprehensive treatment of orthogonal polynomials, hypergeometric functions, Hill's equation, much more. Bibliography. Index. 322pp. 5⅜ x 8½. 65214-9 Pa. $9.95

NUMBER THEORY AND ITS HISTORY, Oystein Ore. Unusually clear, accessible introduction covers counting, properties of numbers, prime numbers, much more. Bibliography. 380pp. 5⅜ x 8½.   65620-9 Pa. $10.95

THE VARIATIONAL PRINCIPLES OF MECHANICS, Cornelius Lanczos. Graduate level coverage of calculus of variations, equations of motion, relativistic mechanics, more. First inexpensive paperbound edition of classic treatise. Index. Bibliography. 418pp. 5⅜ x 8½.   65067-7 Pa. $12.95

MATHEMATICAL TABLES AND FORMULAS, Robert D. Carmichael and Edwin R. Smith. Logarithms, sines, tangents, trig functions, powers, roots, reciprocals, exponential and hyperbolic functions, formulas and theorems. 269pp. 5⅜ x 8½.
60111-0 Pa. $6.95

THEORETICAL PHYSICS, Georg Joos, with Ira M. Freeman. Classic overview covers essential math, mechanics, electromagnetic theory, thermodynamics, quantum mechanics, nuclear physics, other topics. First paperback edition. xxiii + 885pp. 5⅜ x 8½.   65227-0 Pa. $21.95

CHALLENGING MATHEMATICAL PROBLEMS WITH ELEMENTARY SOLUTIONS, A.M. Yaglom and I.M. Yaglom. Over 170 challenging problems on probability theory, combinatorial analysis, points and lines, topology, convex polygons, many other topics. Solutions. Total of 445pp. 5⅜ x 8½. Two-vol. set.
Vol. I: 65536-9 Pa. $7.95
Vol. II: 65537-7 Pa. $7.95

FIFTY CHALLENGING PROBLEMS IN PROBABILITY WITH SOLUTIONS, Frederick Mosteller. Remarkable puzzlers, graded in difficulty, illustrate elementary and advanced aspects of probability. Detailed solutions. 88pp. 5⅜ x 8½.
65355-2 Pa. $4.95

EXPERIMENTS IN TOPOLOGY, Stephen Barr. Classic, lively explanation of one of the byways of mathematics. Klein bottles, Moebius strips, projective planes, map coloring, problem of the Koenigsberg bridges, much more, described with clarity and wit. 43 figures. 210pp. 5⅜ x 8½. 25933-1 Pa. $6.95

RELATIVITY IN ILLUSTRATIONS, Jacob T. Schwartz. Clear nontechnical treatment makes relativity more accessible than ever before. Over 60 drawings illustrate concepts more clearly than text alone. Only high school geometry needed. Bibliography. 128pp. 6⅛ x 9¼. 25965-X Pa. $7.95

AN INTRODUCTION TO ORDINARY DIFFERENTIAL EQUATIONS, Earl A. Coddington. A thorough and systematic first course in elementary differential equations for undergraduates in mathematics and science, with many exercises and problems (with answers). Index. 304pp. 5⅜ x 8½. 65942-9 Pa. $8.95

FOURIER SERIES AND ORTHOGONAL FUNCTIONS, Harry F. Davis. An incisive text combining theory and practical example to introduce Fourier series, orthogonal functions and applications of the Fourier method to boundary-value problems. 570 exercises. Answers and notes. 416pp. 5⅜ x 8½. 65973-9 Pa. $11.95

AN INTRODUCTION TO ALGEBRAIC STRUCTURES, Joseph Landin. Superb self-contained text covers "abstract algebra": sets and numbers, theory of groups, theory of rings, much more. Numerous well-chosen examples, exercises. 247pp. 5⅜ x 8½.
65940-2 Pa. $8.95

STARS AND RELATIVITY, Ya. B. Zel'dovich and I. D. Novikov. Vol. 1 of *Relativistic Astrophysics* by famed Russian scientists. General relativity, properties of matter under astrophysical conditions, stars and stellar systems. Deep physical insights, clear presentation. 1971 edition. References. 544pp. 5⅜ x 8½.
69424-0 Pa. $14.95

Prices subject to change without notice.

Available at your book dealer or write for free Mathematics and Science Catalog to Dept. GI, Dover Publications, Inc., 31 East 2nd St., Mineola, N.Y. 11501. Dover publishes more than 250 books each year on science, elementary and advanced mathematics, biology, music, art, literature, history, social sciences and other areas.